ASTRONOMY FOR BEGINNERS

Astronomy for Beginners

HENRY BRINTON F.R.A.S.

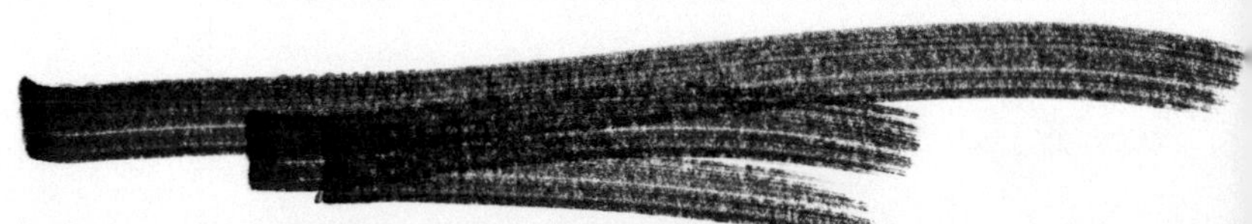

PELHAM BOOKS

First published in Great Britain by
PELHAM BOOKS LTD
52 Bedford Square
London, W.C.1
1970

7207 0393 X

Set and printed in Great Britain by
Tonbridge Printers Ltd, Peach Hall Works, Tonbridge, Kent
in Caledonia ten on twelve point on paper supplied by
P. F. Bingham Ltd, and bound by James Burn
at Esher, Surrey

CONTENTS

ILLUSTRATIONS

FOREWORD

I have tried to cover a great many aspects of a complex subject in a short space. Inevitably this means a certain amount of over-simplification; but that is the necessary price of such a general introduction as I have attempted. If, in the upshot, I have succeeded in awakening the interest of the beginner, and given him a fair picture of the whole field, I shall have succeeded in my object, and make no apology for shortcomings which could not have been avoided.

One apology I do make: There is a good deal of repetition, and I have deliberately sacrificed elegance of writing for ease of understanding. When dealing with such a subject under separate headings, either one must repeat what has been said before, or else subject the reader to constant back-references, which interrupt both the free flow of reading and the process of comprehension.

I must record my grateful thanks to Iain Nicolson, who read the book in typescript, and made many valuable suggestions. Also to Dr J. G. Porter, Dr John Braybon and Dr K. Dyer, who patiently and kindly helped me. I need hardly say that none of them is responsible for any of my shortcomings or faults.

As I have done before, and hope that I shall live long enough to do many times again, I must thank Patrick Moore for his generous help and advice. Astronomy owes him a great debt, and I am happy to pay my individual tribute. I should also like to thank personally those who have kindly allowed me to reproduce their photographs. One of the most pleasant aspects of being an amateur astronomer is the sense of belonging to a group of friends and colleagues who are always ready to share their interests and the products of their work.

CHAPTER I

The Beginnings of Astronomy

Astronomy is the oldest of the sciences, unless one is prepared
to call primitive medicine a science, with its mixture of experience and magic.

Astronomy had much the same basis in its origins. The chief
incentive was a desire to forecast the future. It was a branch
of fortune-telling, a service which it still performs today for the
more silly and credulous. It had also a practical use, which
was almost in the nature of a by-product. It gave the necessary
information on which to found a calendar, which is an essential
for a society which has graduated from food gathering to food
growing. One must know when to plant the crops.

The need to know this was particularly necessary in the early
Egyptian civilization, where the sudden flooding of the Nile
was the foundation of a rich agricultural system. Constant observation showed that this life-giving bounty of the river followed shortly after the bright star Sirius first became just visible
in the early dawn.

Primitive people, and even those of early civilizations, tend
to equate the constant following of one event by another with
a causal connexion between the two. Laymen may smile at this
simple association; philosophers will not. At any rate it is easy
to see that from observing the events in the heavens as the
signal for other events on the earth it was an easy and natural
step to employing astronomy as a means of forecasting. This
was fortunate for later generations. Just as the passion for
alchemy laid the foundations of chemistry, so fortune-telling
gave a great impetus to the science of astronomy, which went
far beyond the need for developing a calendar. Egyptians and

Babylonians, though dependent on the naked eye and the most primitive instruments, developed observational astronomy to a remarkable degree of efficiency. Later, the Greeks added a good deal of theory, and we can use very ancient observations to check modern theories; for instance, the slowing of the earth's rotation.

To however high a degree observational astronomy is developed, the science will not progress very far without two essential facts, both of which may now seem self-evident, but each of which required a terrific effort of imagination to conceive. The first is that the earth is round, and the second that the heavenly bodies are at immense and different distances away from us.

That the earth is a globe is only credible if one also understands the principle of gravitation; even so, acceptance of the fact also involves ideas about space which many laymen find difficult. If the earth is 'floating' in space, what do we mean by space? Has it an end? If so, what lies beyond it? Or how can it go on for ever? One of the most modern answers – that space is finite but boundless – may satisfy mathematical requirements, but is not much use as an aid to visual conception. It would have been meaningless to early peoples.

They had a superficially easier answer. The heavenly bodies were attached to a crystal sphere which revolved about the earth. Beyond the crystal sphere was heaven. There is a delightful drawing of a man peeping through the sphere at the cog-wheels which drive the stars in their courses.* The idea of 'heaven', however, as being beyond the stars does not really help a great deal, since one is left with the query: 'What lies beyond Heaven?'

Putting these unanswerable questions aside, many of the more-intelligent Greeks, believing the evidence of their eyes, accepted the fact that the earth is a globe. It is only necessary to watch how a ship disappears over the horizon in any direction to be made aware of the truth. More convincingly still, one can observe how the Pole Star, Polaris, gets higher as one goes north and declines as one goes south. Only the belief that the earth is round will account for the patent facts.

The Alexandrian Greek, Eratosthenes, went further and actually measured the size of the globe by observing the length

* See top photograph facing page 33 where the picture is reproduced on my run-off shed.

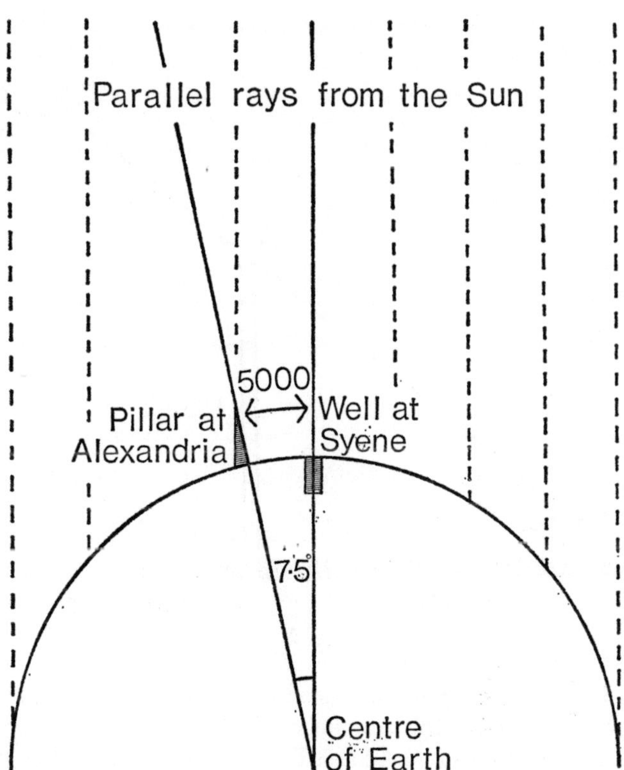

Fig. 1. At Syene the sun shone vertically down a well.
At Alexandria the shadow of a pillar was 7½° from the
vertical. The distance was 5000 stadia. He therefore
calculated that the circumference of the earth was
360/7½ multiplied by 5000 stadia

of shadows of the sun from positions north and south of one
another. He seems to have arrived at a very accurate estimate;
one a good deal better than the figure which Columbus worked
on, and which caused him to mistake the West Indies for the
East.

Since there is nothing like doing an experiment oneself to
understand it, it is worth while invoking the assistance of a
colleague and measuring the size of the earth in the same way
as Eratosthenes. All one needs is a friend a few hundred miles
north or south of one's own position, and a sunny day. With a

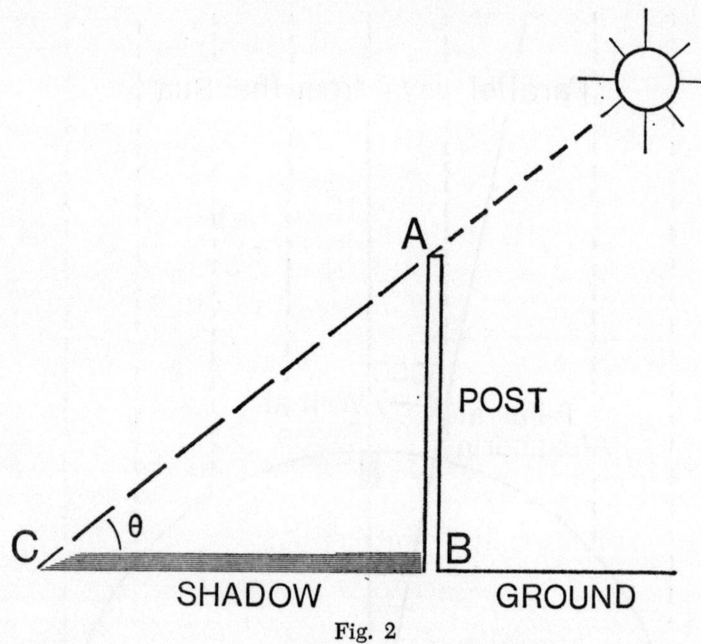

Fig. 2

plumb-line, set up a post on a level piece of ground and measure
the shadow cast by the sun near midday. Keep on measuring
it, until a minimum is reached, and the shadow begins to
lengthen again. Then take the minimum length and work out
the elevation of the sun above the horizon. This can be done by
looking up the tangent of the angle θ made by the height of the
post divided by the length of the shadow. Or, if even log
tables are a sealed mystery, draw the post and shadow to
scale, and measure the angle with a protractor.

If the colleague who lives north or south goes through the
same procedure the same day, you will be able to know the
difference of the sun's maximum altitude between the two positions. Since the sun is so far away that its rays can be taken as
parallel, the most elementary geometry shows that the difference
in the elevation of the sun from the two positions is exactly
equal to the angle subtended at the centre of the earth by them
– as the diagram will make clear.

Let us suppose that the difference is 6°. It follows that the
angular distance apart of the two places is 6 divided by 360

or 1/60th of the earth's circumference. If they are 400 miles north and south of one another, the earth must be 400 x 60 miles in circumference. That is to say, 24,000 miles; which is near enough.

The idea that all the heavenly bodies are the same distance from us, and fixed to a crystal sphere, does not bear much thinking about. In the first place, it is only the stars which keep the same relative positions. The sun and moon, and the planets wander about over the heavenly sphere. Indeed, the name planet is derived from the word meaning wanderer. Furthermore, observation of an eclipse of the sun shows that the moon passes between the earth and the sun, and so must be closer to us. In the same way, the moon frequently passes between us and the various stars, occulting them. One Greek, with the assistance of some ingenious logic, actually worked out the relative distances from us of the moon and the sun, and arrived at an answer which bore some relation to the truth.

The one thing at that period which no one seriously questioned was that the earth was the centre round which the various heavenly bodies revolved. This is thought to have been very naïve; but it is a perfectly sensible way of looking at the question. It is possible to build an orrery – that is, a mechanical working model of the celestial system – and fix the earth or any of the other heavenly bodies so that the others revolve about whichever body is chosen. To take the earth as the centre is merely a trifle egocentric. The real objection to the ancient cosmology is that the taking of the earth as the fixed point makes a description of the movements of the heavenly bodies extremely complex.

The complexity was increased by an attitude of the Greeks, a people who have never been surpassed in their genius for abstract thought. Like a famous detective of fiction, they thought that all problems could be solved by sitting with one's eyes shut, and using the 'little grey cells'. Unfortunately, the little grey cells told them that the perfect figure was the circle. It therefore followed, though the logic is less than perfectly lucid, that the heavenly bodies moved in perfect circles. To fit the observed motions into this scheme, it was necessary to invoke the idea of epicycles – that is, circles whose centres themselves revolve about the circumferences of other circles. By

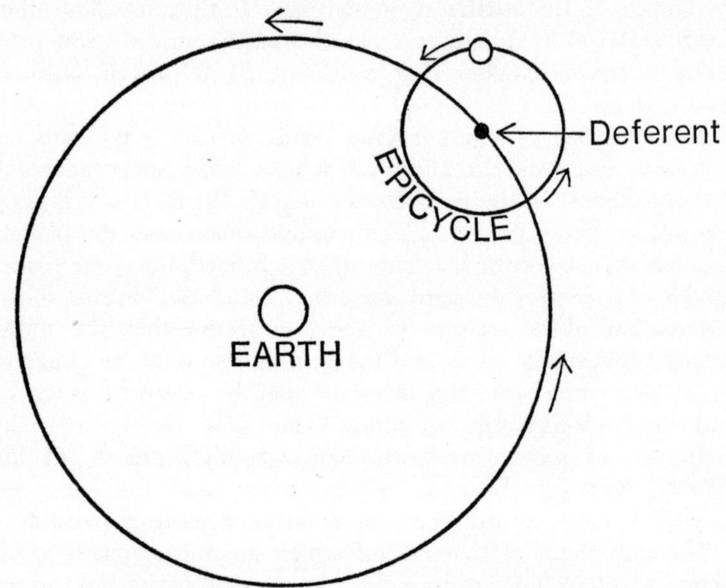

Fig. 3.   The body revolves round the deferent, which itself
moves round the larger circle or epicycle

using this concept, it was possible to work out a scheme which
accounted for the observed motions, within the limits of the
observational errors of the period, and with the earth at the
centre of the universe.

Ptolemy (A.D. 120–180) collected and collated the then known
facts and ideas and for centuries his work represented the sum
of astronomical knowledge, at any rate in the Western world.
It was not until the end of the medieval period that science
began to move again in Western Europe. But it is a common
fallacy to believe that nothing happened anywhere in the interval. From the earliest period of civilization in Sumeria to the
present day, and leaving the isolated culture of China out of
account, there has been an unbroken succession of civilized
nations. Whilst most of Europe endured the darkness of ignorance and near barbarism, a high civilization flourished in
the Middle East and India. Even in Europe itself, Byzantine
culture endured in some splendour until the light of learning
began slowly to filter back to the rest of Europe. Mostly it

came by Sicily and Spain, especially through Spain, brought by the so-called Moors, mainly Arabs, with some Berbers, who had conquered the great seats of learning in Baghdad and Damascus, and absorbed their culture, as a sponge takes up water in a bath.

Most of the ancient Greek works, together with many Latin ones, came back to us in Arabic translations; but the 'Moorish' invaders brought more than that. Far more. The old ideas and observations were a revelation to the backward people amongst whom they now spread; but, alone, they would not have been a great deal of use. Astronomy is a mathematical science; and without the tool of mathematics it cannot progress far. Even the glory of Greek thought could not carry science far without a system of numbers capable of easy manipulation. If one takes the Roman system, and tries to multiply MCDLII by MDLVII, one becomes aware of the hopelessness of the task. These peoples did not even possess the most basic of all mathematical tools – the zero. It is to the Indians that we owe the priceless gift of a working system of numerals, which the Arabs brought with them.

Once knowledge returned to Western Europe, it flourished quickly. So far as astronomy was concerned, the renewed access to knowledge was coupled with a new and practical need for further advances. Perhaps it is significant that the great voyage of Columbus was started in the year in which the last of the Moorish kingdoms, Granada, fell to the Spanish armies of Ferdinand and Isabella. Sailing around the great expanse of the oceans, where there are no landmarks, calls for some system of finding one's way, and astronomy is the only candidate for the post.

In its simplest form, and given that one knows the circumference of the earth, nothing is easier than to find one's distance north or south of the equator. Indeed, some quite effective way-finding can be done without even knowing the size of the earth, given only that one knows the latitude of the desired destination. The method is really only a variant of the experiment of Eratosthenes, to which we have already referred.

At the time of the equinox, when the sun is at its zenith, that is immediately overhead, at the equator, its distance from the zenith at some other point is the latitude of that point. An allowance can be made at other times of the year to do the same sum. A very simple instrument, called an astrolabe, will measure the height of the sun above the horizon with fair accuracy. It consists simply of a circle marked in degrees of wood or metal, with a rotating arm at the centre, along which one sights the lower limb of the sun. It is easy enough to make one for oneself, should one so desire. For an actual measurement one adds a quarter of a degree for half the sun's diameter, since the elevation should be measured from the centre rather than the lower limb.

On first discovering a new place, an explorer would measure its latitude with an astrolabe, or some similar instrument. When he wished to return there, say from England to Jamaica, all he would have to do would be to set off into the Atlantic in a south-westerly direction, so that he would arrive at the latitude of Jamaica well to the east of the island. He could then sail along the latitude to the west until, in due course, he arrived in Jamaica.

As exploration spread wider and wider, and journeys to far parts became more common, a pressing need developed to have better and better astronomical information. Latitude sailing was all right in its way, but it would clearly be much more satisfactory to be able also to measure longitude. In fact, it can only be done easily and accurately by having accurate clocks; but these were not developed until well into the eighteenth century. In the meanwhile, ingenious ways were found of getting some sort of longitude from the movement of the moon among the stars; a need which necessitated fuller and more accurate star maps.

In the meanwhile, the age-old idea that the earth was the centre of the universe was beginning to be shaken at last. Copernicus (1475–1543), a Pole by birth, devised a new system, which though it still made use of the concept of epicycles, put the sun at the centre. Strangely, at least to modern ears, the idea that the earth, which is God's creation, should not be at the centre of all things, was offensive to the Church. Any suggestion of the kind was liable to lead to nasty consequences. Copernicus avoided trouble, partly by wrapping up his thesis; but, later on, Giordano Bruno (1550–1600) was burned at the stake, partly because he advocated the Copernican theory.

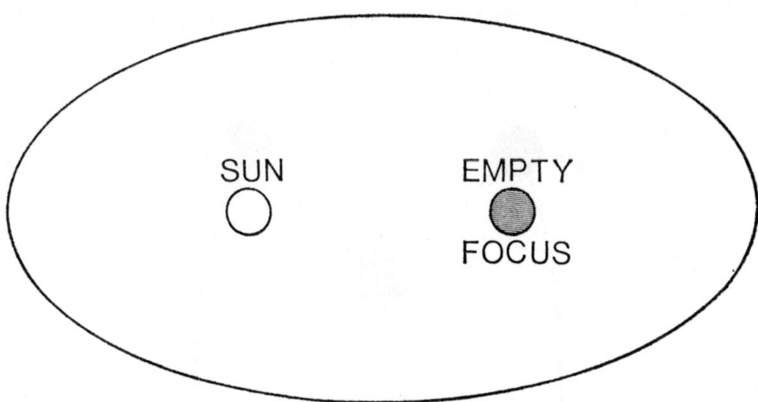

Fig. 4.   An ellipse has two foci. For planetary movements the sun
occupies one, the other is empty

Science, however, cannot be stifled. A quaint observational
genius, Tycho Brahe (1546–1601), made a series of extraordin-
arily accurate observations of planetary movements, in spite of
the crude nature of his instruments. It was then left for the
great Johann Kepler (1571–1630) to systematize the facts into
a theory. It was Kepler who broke astronomy loose from the
bonds of the old Greek concept that all celestial movements are
in circles. He showed that, on the contrary, the planets move in
ellipses round the sun, with the sun at one focus of the ellipse.
He showed too that, as would be required by the later gravitational theory of Newton, the speed in orbit of the planet is
slowest when furthest from the sun, and fastest as it is nearest.
As he stated the law, 'the line joining the planet to the sun
sweeps out equal areas in equal periods of time'. (see Fig. 5)

The moment was now come for perhaps the greatest single
stride in the history of astronomy – the use of the telescope.
Crude lenses had long been known; but at the very start of the
seventeenth century a Dutchman found a way to combine lenses
into a crude telescope. Galileo (1564–1642) heard of this strange
new device and proceeded to make one for himself. In a breathtaking moment of history, he turned his crude instrument upon
the planets; and in that moment modern astronomy was born.

Galileo was famous in other fields besides astronomy – he was,
for instance, the effective founder of the skill of experimental

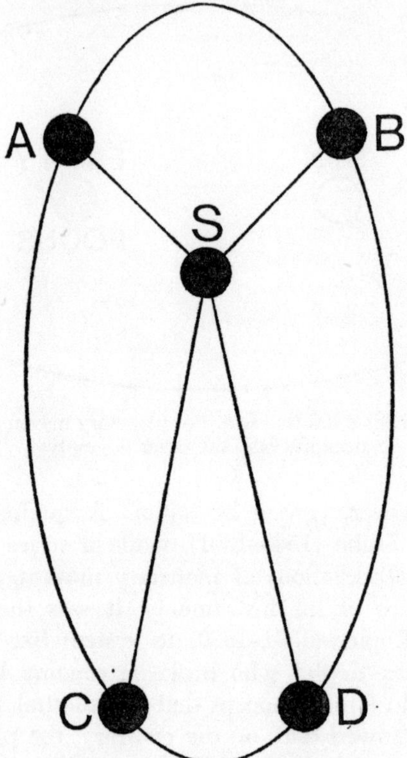

Fig. 5. The segment ASB has the same area as
CSD; but the part of the ellipse from A to B is
longer than that from D to C, so the planet moves
faster at the position nearest to the sun (S)

mechanics – but he has come down in history particularly for
two historic events. When he used his telescope on the heavens,
he made a number of classic observations. He observed the
mountains and craters on the moon, and he remarked upon the
four satellites, or 'moons' of Jupiter. It was, however, his discovery that the planet Venus has phases like the moon which
was destined to strike the death-blow to the ancient and vener-
able belief that the earth was the centre of all things.

It is often said that the discovery of the phases of Venus
'proved' that the earth goes round the sun, and not vice versa,
but this is not strictly true. As we have seen, we can consider

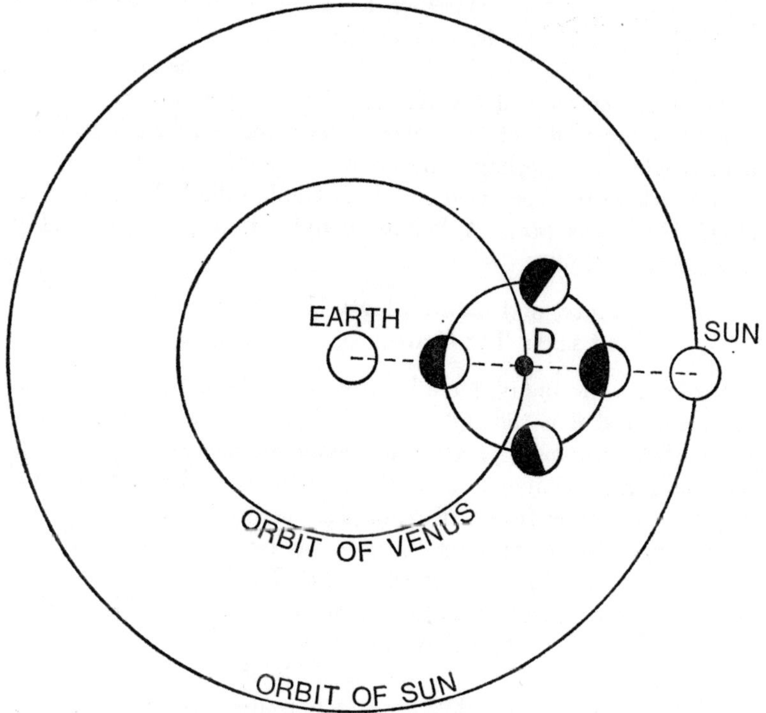

Fig. 6. The Ptolemaic theory of Venus. The line joining the earth to the sun, and passing through the defferent D was always a straight line. Therefore Venus would never be seen except as a crescent

the matter either way, if we are prepared to accept a great complexity of description and some absurdities about the star field. What was true and important was that the Ptolemaic theory would not accommodate a Venus which exhibits phases, whilst a simple heliocentric (i.e. sun-centred) system would.

Galileo's second claim to a place in history derives from his championship of the heliocentric theory, against the bitter opposition of the Church. Times had changed to the extent that he escaped the fate of Bruno, though he was imprisoned and compelled to recant in public. The story runs, though it is probably untrue, that after speaking the words of recantation, he muttered under his breath, though still audibly: 'Eppur si movere'

– 'But it *does* move.' Whether he actually said it or not, he undoubtedly thought it, and educated men everywhere came to accept the view, bowing to the logic of hard facts.

With the coming of the telescope and the acceptance of the modern conception of the solar system, the way was clear for a man who was, perhaps, the greatest scientific genius who has ever lived. In the very year in which Galileo died, Isaac Newton (1642–1727) was born. A famous couplet is only a pardonably exaggerated tribute:

> Nature, and nature's laws, lay wrapt in night.
> God said: 'Let Newton be' and all was light.

Books, by the many, have been written about this genius and his work, and it is difficult to select from such a treasure-house for a short summary, and still convey an adequate picture of Newton's achievements. One thing, however, does stand out, even in such a profusion of triumphs. Newton's theory of gravitation, whereby he brought the heavens into line with terrestial laws, has stood through the centuries. Today, it is limited by the theories of another all-time genius, Albert Einstein; but it has been extended rather than superseded.

The story of Newton and the apple in the orchard is one of the few bits of nursery history which is probably true, though the moral is generally distorted. It is often said that the falling apple suggested to Newton the idea of gravitation; but this is not true. The phenomenon of gravitation was well known. What Newton thought of in the orchard was the, to us, obvious, but to his generation startling, idea that the force of gravitation which made the apple fall to the earth was the same force which kept the heavenly bodies moving in their appointed courses.

To the earth-bound human senses, it seems obvious that the tendency of a moving body is to come to rest, unless some force continues to propel it; yet this conception is wholly wrong. A moving body will continue to move in a straight line, and at the same speed, unless and until some external force is applied to cause it to alter its direction or speed. The point is simply that, in our everyday experience, the force of gravitation, together with the friction of bodies rubbing together or meeting the resistance of the air, always does cause moving bodies to

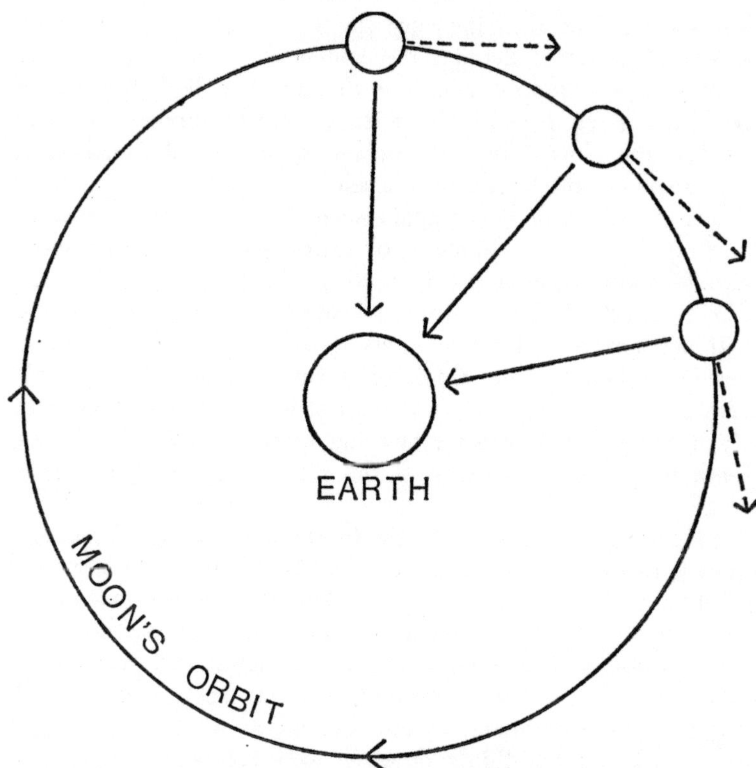

Fig. 7. At a part of its orbit, the moon would fly off at a tangent (dotted lines) if it were not held back by the attraction of the earth

alter their motions and slow down. Out in space where bodies are isolated and there is no air to restrict their movement, bodies move freely in lines which would be straight but for their mutual gravitational attraction.

There was one easy way to make a first check on this revolutionary concept. The moon moves round the earth. (Actually, as we shall see later, they move in fact around their common centre of gravity; but for the moment we can neglect this refinement.) It goes in orbit around us, rather than flying off into space in a straight line because it is chained to us by the force of the mutual gravitational field. For this to be true, it follows that the tendency of the moon to fly off must exactly balance the pull of the earth. Anyone can check for himself that this is true, if he uses the very simplest mathematics. The calculations is shown in detail in the Appendix.

All genius is simple, and it is delightful to find that one of the great steps forward in science can be understood and checked for himself by the average schoolboy of fourteen. If one battles out the figures for oneself, the mysterious calculations of astronomers will never again seem to be so mysterious. So much of the basic arithmetic of astronomy requires no more than 'O' level Maths that it is worth proving it to oneself.

One aspect of this charmingly simple sum may puzzle the beginner, and must be elucidated, if we are to follow Newton's reasoning. We humans think of a force in terms of pounds weight, but in astronomy deal with it in terms of acceleration, which at first sight seems something quite different. To understand, we must look briefly at the relationship between mass, weight and inertia.

Weight, as soon as we begin to think in terms of modern space travel, is an elusive concept. How much would a pound of sugar weigh on the moon? To find the answer, we might take a pound of sugar and a pair of scales to the moon and weigh it there. The answer would be that it still weighed a pound. If, however, we also took a spring scale with us and repeated the experiment, lo and behold, it would only weigh about two and two-thirds ounces! And yet both instruments would be 'right'. This seeming contradiction is because 'weight' is a local and not a universal concept. Weight belongs as a fixed scale to any one gravitational field, and is different in another. There is one scale on earth, and another on the moon, just as there are other weight scales for other planets.

Fortunately, there is an underlying, and fixed, quantity to every scale. This is the quantity known as inertia, which is a kind of natural laziness inherent in all matter, and which does not vary with varying gravitational fields. To make matter move, or stop it moving, or to change its velocity or direction, a force has to be applied. If we put a billiard ball on a sheet of smooth ice, where there is hardly any friction, we still have to exert a force to make the ball move. If it is moving, we have to apply a force to stop it, or a different force to change its direction of movement. The force, as simple experiments would show, is in proportion to the weight of the billiard ball; if we substituted a smooth cannon-ball for the billiard ball, we should need a great deal more force applied to make it move at the same rate as the billiard ball.

If we took the ice and the balls to the moon, where the gravitation is less, or out into space, where the gravitation is negligible, we should have to exert exactly the same forces to make the balls move at the same speeds. Since the weights of the balls would be different in the different gravitational fields (in space it would be substantially zero), we cannot measure inertia in terms of pounds or kilograms, which are local concepts. We use instead the principle of acceleration, which is universal. Acceleration is merely the rate of change of velocity. When we push a billiard ball on ice, it may seem to be still at one moment and moving at a fixed velocity the next. But this is not so, as we should find out with a very large and very heavy cannon-ball. It would start slowly, and continue to move faster and faster for as long as we went on pushing. That is, it would accelerate. Acceleration is measured in feet per second per second. Mass is a measure of a body's inertia, it is measured in terms of how much acceleration would be given to that body by a given force. The greater the mass, the greater the force needed to move it at a given speed.

Now, Newton stated his gravitational theory in terms of mass. He laid down that the gravitational attraction between two bodies is the product of their masses divided by the square of their distance apart. When we weigh a body, we are measuring the gravitational attraction upon it of the earth. (Strictly speaking, the measure of their mutual attraction, though the object being weighed is generally so trivial in comparison with the earth that we ignore its mass.) Clearly, the greater the mass of the object, the more it will be attracted by the earth, and so the more we shall say that it weighs. Weight, then is governed by its inertial mass and the strength of the gravitational field in which it happens to lie. To quote a happy phrase of Patrick Moore: Gravitation is the force which gives mass weight. The mass is universal: the weight of a characteristic of the mass and the local gravitational field.

One constant is missing before Newton's theory of gravitation can be applied to the universe with full effect. Newton did not state, and did not know, the *amount* by which masses attract one another. This could only be found by actual measurement; a feat of great delicacy. It is worth diverting our short-historical survey a little to see how the problem was solved by Henry Cavendish (1731–1810). The experiment can be done in either of two forms. As Cavendish did it, it took the form of suspending heavy spherical weights near to one another, and measuring, by direct measurement, how much they attracted one another.

The alternative is worth detailing, because it can be seen that what one is doing is to 'weigh' the earth itself! If one takes an ordinary, and very sensitive, balance and weights it with precise evenness on both sides, it will be level, because the earth is attracting both sets of weights equally. If underneath one pan one now places a heavy weight of known mass at a precise distance below the pan, this mass will add its attraction to that of the earth on that side of the scale, and additional weights will have to be added to the other side to restore the balance. The amount of weight necessary to add, when divided by the square of the distance from the centre of the added mass, can then be related to the pull of the earth, divided by the square of the distance of its centre from the weights in the pan, so giving the mass of the earth. As a matter of interest, the best modern figure is 5.976×10^{27} grammes. More importantly, the universal constant of gravitation, written usually as big G – that is the mutual attraction between two masses – is 6.670 $\times 10^{-8}$ g^{-1} cm^3 s^{-2}. It is remarkable how close Cavendish, with his simple equipment, got to the correct figure.

To return, after this somewhat long excursion into explanations, to the work of Newton. Though his great contribution to astronomy was the proof that the heavenly bodies are subject to exact and comprehensible laws of mutual attraction, he made many more discoveries, each of which would have given him a lasting place in the hall of fame. It was he, principally, who developed the mathematical method of calculus, which did as much to make calculation possible as the computer has done in our age. Unfortunately, Leibnitz (1646–1716) worked along the same lines and there was a silly squabble about who was the pioneer. (To be fair, we must admit that Leibnitz's methods of working out calculus were more convenient than Newton's; but this is beside the point.)

In another field, too, Newton was pre-eminent: that of the

study of light. In the practical field, he devized what is still known as the Newtonian reflecting telescope, which will be described in Chapter IX. More fundamentally, he dealt with refraction, and showed that white light, by refracting it through a prism, could be split up into its component colours: the colours of the rainbow – Red, Orange, Yellow, Green, Blue, Indigo, Violet. (Mnemonic: Richard of York Gave Battle In Vain.) His simple and elegant experiment was the basis on which modern spectroscopy, one of the most powerful tools of astronomy, is founded.

The age of Newton witnessed an astonishing growth of science in many fields besides that of astronomy. King Charles II, who earned fame as an eager, rather than a discriminating lover, and also as a slippery, if adroit, politician, was in addition, a great patron of the sciences. It was he who founded the Royal Society, membership of which still represents the pinnacle of the scientist's ambition. He also founded the Royal Observatory, which, though it has had to move from Greenwich to Herstmonceux Castle remains the world centre of astronomy.

The period is studded with the names of celebrated astronomers, whose observations and theories laid a lasting and solid foundation on which the later generations have built the skills and knowledge which is freeing man from the prison of his little world, helping him, literally, to reach out toward the stars. Most of this work, especially since much of it was endowed for the purpose of forwarding the practical advance of navigation, was concerned with mapping, observing and classifying. Discoveries often followed unexpectedly from attempts to account for observations, or efforts to measure some unknown quantity, as when Bradley, the third Astronomer Royal, discovered the aberration of light when he was trying to measure a stellar parallax.

Any attempt to try to follow the history of the science from this period would require a book to itself, or would lead to a confusing order of explanation. The great steps had been taken, and the story can best be taken up under separate subject headings. But it is impossible to overestimate the magnitude of the achievements of these early pioneers.

Perhaps the most important advance of all had taken place silently and unobserved: the scientific method itself had been evolved. The early Greeks had sought knowledge by a process of pure reason, almost entirely divorced from measurement or observation. The medieval schoolmen had moved backwards. To them all knowledge was already contained in the scriptures and the works of Aristotle. If one wanted to find something new, one merely sought to sift these authentic canons more thoroughly. It was an enormous stride from this mixture of pure logic and holy writ to the idea of experiment and observation. To the idea that if one wanted to know how a stone fell, one did not argue from first principle, or study the works of the Greek philosophers. One dropped a stone over the edge of the leaning tower of Pisa, and observed how it fell, even if Galileo never did the particular experiment with which he is credited.

The second victory of wisdom over habit, was when the geocentric theory was abandoned, and man degraded his own importance in favour of the pursuit of truth for its own sake, irrespective of where it led him. The third and greatest milestone was when Newton took his yardstick into the heavens, and subjected them to the rule of scientific law.

From this time onward, astronomy became a complete, if sometimes erratic, branch of science as a whole.

CHAPTER II

The Physics of Astronomy

In dealing with modern physics, even in a simplified form, one has, unfortunately, to leave the ordinary ideas of commonsense behind. At least the exercise should make one sympathize with the man of long ago, who had to get used to the idea that the earth was round! Lord Kelvin, himself a great scientist, once remarked that he could not understand anything of which he could not make a working model. He would have had a difficult time of it if he had survived a little longer.

The best place to start a survey of the basic physics of astronomy is with the nature of matter itself. As our ideas and knowledge progressed, the nature of matter, for a long time, became gradually simpler and easier to comprehend. The old Greek idea was that matter consisted of 'substance', the underlying stuff of matter, which was itself inapprehensible. To it were attached a series of qualities, or accidents, such as colour, taste, smell, etc. The 'science' of alchemy derived from this concept. By changing the accidents, one could change the nature of an object. If, for instance, one could combine the colour of sulphur with the heaviness of mercury, one might make gold. Just as fortune telling founded the science of astronomy, so alchemy started the science of chemistry.

Gradually it became clear that this picture would not do. A first and great simplification took place when the idea of elements took shape. Matter became divided into ninety-two separate and fundamentally different elements. To begin with, all ninety-two were not known; but that was the sum total of immutable basic substances. The elements might be mixed together, either loosely, so that they could be separated again mechanically, or tightly in chemical compounds, of which there were almost an innumerable variety. However much one tried to sort matter out into its constituent parts, one came eventually to the element, which was indestructible and would always retain its individuality.

This comparative simplicity was capable of further simplification. The smallest and simplest form of matter is the atom; and this consisted essentially of only two components: the proton and the electron. All matter was built up of these two building blocks. All protons and electrons were precisely alike, and the different elements got their individualities from the number of elementary particles they contained.

There was a slight complication in that there were also elementary particles called neutrons; but these could be thought of, roughly, as a mere combination of an electron and a proton.

This was the stage at which our ideas about matter were at their most delightfully simple and comprehensible. Unfortunately, and rather unexpectedly, the picture became complicated again. At a fairly early stage a variety of electron turned up having a different electric charge, and called a positron; but this was not a very serious bar to understanding. There is also a chargeless, particle with no mass, known as the neutrino. Then, in the 1940s, there appeared the particle called the meson, which was in between the electron and the proton in mass. From that moment onwards, a whole series of new elementary particles has shown up, and the search goes on. At the time of writing, the hunt is up for a mysterious and elusive entity: the quark, which is demanded by theory and undiscovered by experiment.

Fortunately, for our purposes, we can neglect these modern refinements, and stick to the simple model of Niels Bohr. The picture is incomplete and out of date; but it will suffice for an understanding of the range of phenomena with which we shall be dealing.

All atoms consist of a nucleus, around which revolve one or more electrons. The nucleus itself consists of one or more protons, together generally with one or more neutrons. The proton is small and heavy and has a positive electrical charge. The electron is large and light, and has a negative charge. The neutrons, which may be present in the nucleus, have no electrical charge.

The chemical behaviour of an element is decided by the number of positive charges on the nucleus: that is to say the number of protons in it, which will be balanced by an equal number of orbiting electrons. The atomic weight of an element, that is the weight of one atom, is, naturally, determined by the number of protons and neutrons plus the lesser weight of the electrons.

The simplest of all forms of matter is a single proton. If it has a circling electron, it is an ordinary atom of ordinary hydrogen. In this sense, we can say that all matter is built up of hydrogen. Even without the electron, the proton will still be hydrogen, since all elements get one or more of their electrons knocked off from time to time.

The next element in order of weight is helium, which has two protons and two neutrons in the nucleus and, appropriately, two orbiting electrons. The question naturally arises of what happens if we get a nucleus of one proton and one neutron? The answer is that we have what is called an isotope of hydrogen. Since the nature of an element is dictated by the positive charge on the nucleus, it follows that such an atom must behave like hydrogen, but will be almost exactly twice as heavy. Such an isotope is called heavy hydrogen (or deuterium). If two such atoms are joined to one of oxygen, the result will be what is called heavy water.

All elements may have isotopes, being merely atoms of the element which have more or less than their proper number of neutrons in the nucleus. Isotopes tend to be unstable. Isotopes of heavy elements tend to absorb or eject one or more neutrons, and split up into two elements which are lighter. We can now understand what used to puzzle schoolboys half a century ago; why the number of elements is limited to the odd number of ninety-two. There is, strictly speaking, no such arbitrary limit. The fact is that any element heavier than uranium, which has ninety-two protons in its nucleus, is so unstable that it cannot survive in nature. In the process of playing about with atomic energy, we have made so-called trans-uranic elements, such as plutonium, with positive charge in the nucleus greater than ninety-two.

Naturally-occurring unstable elements break up over a period of time, as does, for instance, radium; but the break-up may be slow enough for the elements to survive long enough to be still present in our natural environment. We measure the rate of break up in terms of a half-life; that is, the period during which half the atoms in any given sample will break up. We cannot, at the moment, say exactly why any particular atom should choose any given instant to break up; but we do know that the rate at which they do it is constant for any given element. A half-life may be anything from a small fraction of a second to many thousand years. Incidentally, long-dead alchemists must be having a quiet chuckle in the never-never land to see their basic idea vindicated, even if their theories were a pack of rubbish.

So far, our picture of the atom is reasonably clear and simple. Lord Kelvin could almost have made a working model, though he would have had difficulties. He could not, for instance, have told us how protons, each having a positive charge, keep together, when the positive charges must repel one another. We come now, however, to a more difficult concept.

We have spoken of the positron; the electron with a positive charge. Of itself, it does not take a lot of comprehending; but its reaction with an electron leads us into strange country. If the two come together, the result is an 'explosion', in which the two particles vanish in a flash of energy. When I was a boy, and got a chemical equation wrong, I was made to write out five hundred times: 'You cannot create matter, and you cannot destroy it.' My punishment was unjust! A positron and an electron do destroy one another. Substantially, they are changed from matter into energy. But not only electrons have their identical twins with an opposite charge; there are protons with negative charges. It is possible to conceive of a different kind of matter – anti-matter, if you like – composed of negative protons and positive electrons. The two kinds of matter would react in the most violent way, disappearing altogether in the form of matter, and transposed into pure energy. Such an explosion would make a hydrogen bomb seem like a child's toy.

This is where we get outside the capacity of the human being to envisage an aspect of nature which is outside his experience. When we think of energy, we think of it in terms of pushes and pulls. That is of something being pushed or pulled. We cannot visualize a force without something being subjected to it; yet we must now think of matter and energy as being inter-

Above Author's 12″ telescope. The large circle is for declination. The smaller, at the top of the stand, is the R.A. circle. The larger and smaller sighting telescopes are seen at the forward end. This photograph was taken by the author with a wooden box and one of his spectacle lenses. *Below* Author's 12″ telescope, with Patrick Moore at the eye-piece. His hand is near the hand adjustment for declination. The revolving head is shown

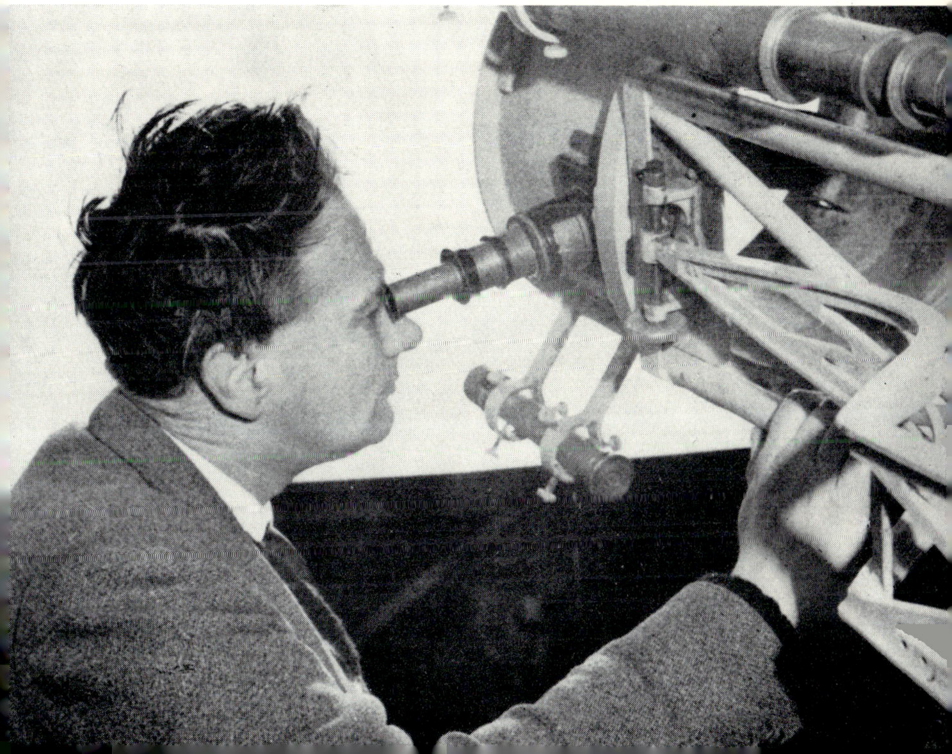

Above Run-off shed for author's 12″ telescope. *Below* Revolving dome on Patrick Moore's 8″ telescope. The glass top turns. Note open shutter

changeable. Matter, we may say, is a mere manifestation of energy: a pattern of energy, or a knot of energy. We can give the concept mathematical expression; but we cannot visualize it.

Einstein, in fact, gave a precise mathematical expression to the relationship between matter and energy. His equation is one of the simplest and most basic in nature: $e=mc^2$. e is the energy in ergs released. m the mass of the matter transmuted in grammes and c the gigantic figure of the speed on light in centimetres per second: 3×10^{10}. The square, therefore, is 9×10^{20}. The amount of matter destroyed, or transmuted to energy, in, for instance, a fission bomb is very small indeed; but the energy released is, because of this enormous figure, very large – as Hiroshima and Nagasaki learned to their sorrow and mankind's shame.

The energy of an atom or fission bomb is easily calculated. One merely needs to know the mass of the original atoms of, say, plutonium, and the mass of the total fission products, and apply Einstein's equation. Atomic fission, however, does not play a conspicuous part in the physics of astronomy. It is the opposite process of atomic fusion, which is the main fuel of the celestial power-house. Two atoms of heavy hydrogen will form in combination an atom of helium, which will have a mass very slightly less than the original two heavy hydrogen atoms. The difference will be the energy released, again calculated in terms of $e=mc^2$. The process is not always quite so simple. It may take the form of the combination of an atom of tritium, that is one proton with two neutrons, and an atom of ordinary hydrogen, or, as in the sun, the process can be more involved still. It is also complicated by the fact that, in normal surroundings, the act of fusion requires an enormous input of energy to start it off.

The sun and the other stars have a main source of energy in this fusion of hydrogen into helium. They start as mere aggregations of hydrogen, or very nearly pure hydrogen, and gradually build up heavier elements, releasing vast quantities of energy, which they radiate away. Without this source of energy, they would burn out in no time at all. One can get a faint idea of the almost inexhaustible energy concerned when we reflect that the sun, which has been in existence for thousands of millions of years, and will go on radiating for thousands of millions of

B

years to come, radiates away about four million tons of matter every *second*.

Perhaps it is as well that hydrogen requires triggering off before it fuses, when we remember that the oceans of the world are largely composed of hydrogen. On the other hand, we have still not tamed hydrogen as a source of power, other than in violent explosions, partly because of the great temperature necessary to start the fusion, and partly because of the difficulty of keeping the hydrogen together once the process has started. When we do tame the process, and it is taking a long time, and many false starts have been made, power should become very cheap indeed, though the cost of distributing it will remain a major expense.

Great though this potential source is, the mind reels before the prospect of being able to transmute complete atoms into energy. That there are still probably processes in nature of which we have so far no clear idea is becoming evident at the time of writing this book. Even if the subject were not too advanced for a beginners' manual, it would not be possible to go far into the most modern physics. The science is in a state of flux, with almost endless possibilities opening up. The study of so-called quasars, and the possibility of what are called neutron stars, suggest that there are sources of energy in the universe almost beyond man's dreaming.

The neutron star, in which it is supposed that the huge gravitational forces of the star have caused the collapse of the structure of the atoms in it, opens up yet another fascinating possibility. As we shall see in a little while, light is itself subject to gravitational forces. There may be stars in which the gravitational field is so great that light from them can never escape at all. That would mean that, even if they existed near to us, we should never see them.

After this dip into the future, it is time to come back to some of the simpler and more basic properties of the atom. We have spoken of electrons revolving or orbiting round the nucleus. When there are a number of electrons, they make up a series of separate rings. Once a ring becomes filled, another ring is started. If the rings are all just completely filled, the resulting element is chemically very stable, and unwilling to combine with the atoms of other elements. Thus the 'Noble' gases: argon, neon, etc., have their outer ring, or shell, filled, and it is impossible, or nearly so, to form them into compounds.

The orbiting electrons, even if, as with hydrogen, there is only one of them, have a choice of a number of sizes of orbit. An atom can absorb energy, which causes its electron to take up an outer orbit, which can then be discharged in radiation by the electron falling back into its original orbit. When an atom has absorbed energy, it is said to be excited. One of the stranger facts of this process is that, though electrons have alternative orbits, they are never in the space between. They can pass from one orbit to another, without traversing the intermediate space.

The energy which is thus radiated or absorbed is in the form of electro-magnetic waves, the most familiar form of which is what we call light. Light is simply that wavelength which our eyes can receive and change into messages to our brains. Light is no different in kind from any other electro-magnetic wave, varying from the excessively short gamma rays, with wavelength only the tiniest fraction of a millimetre, through the longer X-rays to ultra-violet light. Then comes the narrow band of visible light, followed by infra-red, radiant heat, ultra short-wave radio, and so on to the kilometre long waves of the older radio.

One question which has perplexed scientists is the nature of these waves; in particular, the nature of light. Newton took part in the controversy which raged about it. In a sense, the problem has never been solved. There are two ways in which we can imagine light as travelling: either as a collection of particles, like small bullets, or else like waves. If light consists of particles, it is easy enough to see how it can get from one place to another. If, however, it consists of waves, and most of the evidence seemed to suggest that it travelled in waves, then it would need a medium in which to travel. Waves at sea are obviously movements on the water. Sound consists of waves travelling through the air. We can make visible waves for ourselves by suspending a light coil spring, and hitting one end of it. Wind makes waves in a field of corn. But, if light were waves, what medium was it through which it passed in the empty reaches of space?

To meet this difficulty, scientists invented the rather im-

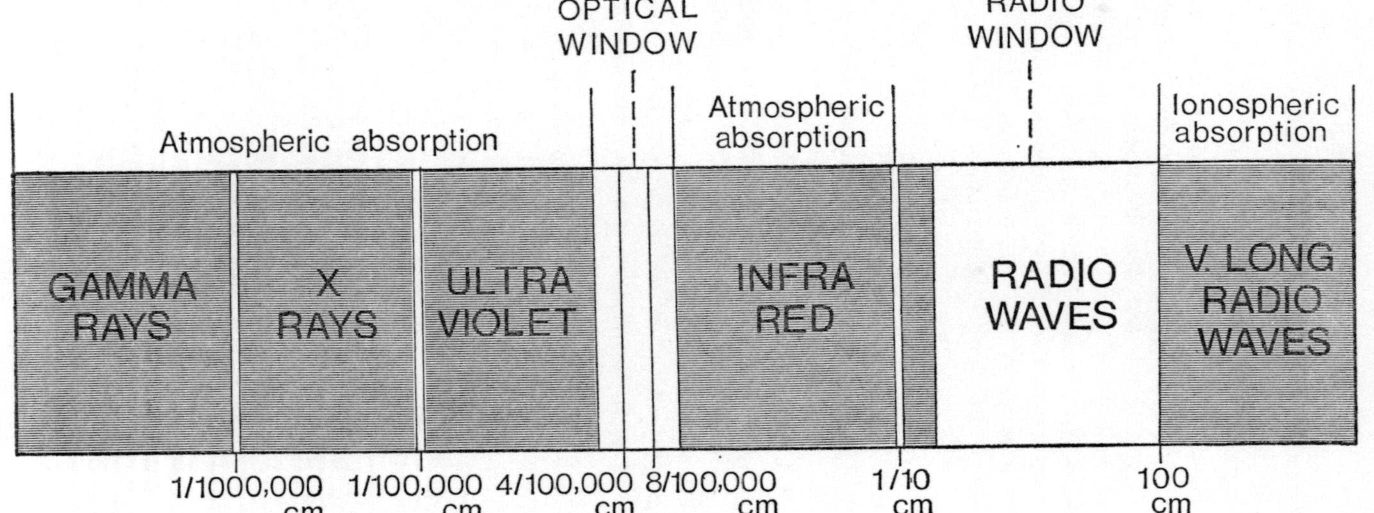

Fig. 8. The electromagnetic spectrum. Only radiations in the narrow band of waves called the optical window are visible to the human eye

plausible idea of the 'luminiferous ether' which was a kind of immaterial matter solely existing to carry electro-magnetic waves. To make it fit the evidence, the ether got more and more complex and less and less credible. Finally, one of the most celebrated scientific experiments of all time was devized to prove the existence of ether, and, incidentally, reveal the true movement of the earth through space. This was the experiment performed in 1887 by Michelson and Morley.

If space were pervaded, as the theory required, by ether, then the heavenly bodies moved through it. The earth would have to be like a ship travelling through a sea of ether. Simple theory tells us that a wave travelling forward along the line of motion, and then reflected back against the line of motion, will take longer than a wave travelling at right-angles to the line of motion, and also reflected back.

The Michelson-Morley experiment consisted of sending out two beams of light at right-angles to one another for exactly equal distances, and then having them reflected back on the opposite course. If the beams were rotated through a circle, then, no matter what the true course of the earth through space, at some stage one beam would be along the true line of movement, and one at right-angles to it. By measuring the time taken by the two journeys, the true speed of the earth would be revealed, and the theory of light as a system of waves would be vindicated.

In one of the most shattering moments of scientific history, the experiment failed to show any difference in time whatsoever! It was repeated again after an interval of six months, when the earth was going in the opposite direction round the sun. The delicacy of the experiment was sufficient to have shown even this difference; but the result was still obstinately negative.

There seemed to be nothing for it but to abandon the ether, which would have caused little distress. Unfortunately, it also appeared to involve jettisoning the wave theory of light itself – which was to throw out the baby with the bath water with a vengeance! There remained the despised theory that light is composed of particles.

An easy way is available to test this theory. If particles are fired from a moving platform, the motion of the platform has to be taken into account. If one fires a revolver with a muzzle velocity of x feet per second in the same direction in which a car is travelling at y feet per second the velocity of the bullet starts as x+y feet per sec.: If one fires it backward, it starts at x−y. Now the satellites of the planet Jupiter move sometimes towards the earth and sometimes away from it, as they revolve round the planet. All that is necessary is to measure the velocities of the light from one satellite moving towards and from one moving away from the earth. On the particular theory the difference in the velocities measured should show a difference of twice the velocities of the satellites. Alas and alack! It shows nothing of the kind.

However we look at the problem, and it is an important one in astronomy, we get the same kinds of self-contradiction. If we study light striking a metal plate, and knocking electrons out of it, we get a result only consistent with the particle theory. On the other hand, we have two important properties of light consistent with waves.

First is the case of light travelling past a very narrow obstruction. If one fires bullets at a screen with a metal post between the gun and the screen, moving the gun in an arc as one fires, there will be a line of bullet holes in the screen, with a blank opposite to the line of the post. On the other hand, if one puts a post in water and causes waves to ride past the post, provided that the waves are of fair wavelength, they will join up again on the far side of the post. If now we shine a beam of light past an upright pin on to a screen, there will be no hard shadow on the screen. The way light behaves in these circumstances is only consistent with the wave theory.

Then there is the phenomenon of interference. Imagine two trains of waves moving in the same direction and of the same wavelengths. If they are in step, that is with wave-crest level with wave-crest, and trough with trough, the result will be a wave train with crests and troughs twice the height of either of the original wave trains. If, however, the crests are level with troughs, and troughs with crests, the waves will cancel out, and the water appear calm. All electro-magnetic waves behave in this way – a property which cannot belong to particles.

We are left with no alternative but to accept that light has the properties of both waves and particles. This is impossible

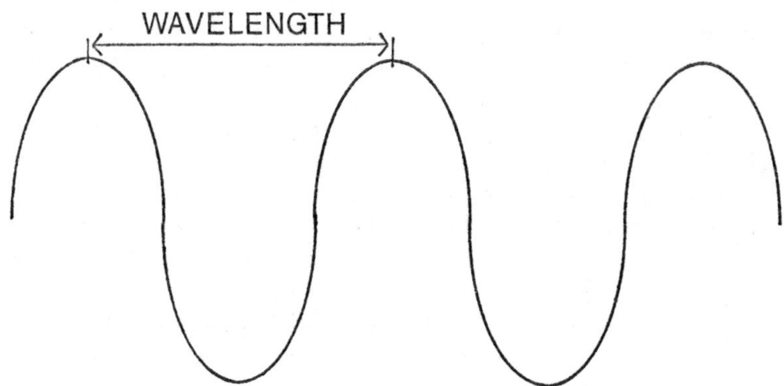

Fig. 9.   A wavelength is measured from crest to crest

to conceive; but a perfectly simple matter to the mathematicians, and we have to accept that physics are ruled by mathematics in a way that allows us to predict and explain; but not to picture in our minds. We can, then, think of light either as small bullets, which are called photons, or else as trains of waves with specific wavelengths, which dictate the colour of visible light, and the nature of other forms of radiation. Violet has the shortest wavelength and red the longest, with the other colours ranked in the order of the rainbow. It is sometimes said that we can think of light as discrete photons, and the wave nature as being the probability of finding a photon at a given place at a given instant. In other words, the waves are waves of probability.

It only remains to be added, and perhaps one should anticipate the event, that, though we have pictured electrons as particles, they can be made to behave like waves. The microcosmic world is very odd indeed, and we have only touched upon the easiest of the oddities.

Another oddity which ought to be apparent from what has been written is the constant speed of light through a vacuum. Let us look at it this way. Suppose we have a stationary source of light and two objects moving towards or away from this source at different velocities with observers on each of the moving bodies. If they both measure the velocity of the light which they receive, they will get exactly equal values: almost precisely 300,000 kilometres per second. Light, in fact, is always travelling at the same velocity irrespective of the velocity of either the source or the observer. The velocity of light is a constant in a world of relativity.

Though the velocity of the light, however, is constant, the relative movements of the source and observer make an important difference in how the light is perceived. If observer and source are moving towards one another the wavelength will appear shorter and so the light will be bluer. If they are moving apart, it will be reddened. This phenomenon is known as the Doppler effect, and is the most widely used tool of measuring in astronomy. Indeed, some conclusions are drawn from the reddening of light which are so fundamental to our theories about the universe that one wishes that the ground were stronger.

The Doppler effect is quite easy to understand. We get a similar effect if a car goes tearing by us, blowing its horn as it passes. Whilst the car is approaching, the note of the horn is unnaturally high, and it goes unnaturally low the moment the car is past. The sound is made by the diaphragm of the horn, which first compresses and then thins the air in front of it. These alternating compressions and thinnings of the air pass through it in a train of waves, and transfer the same wave pattern to the human ear. The shorter the wavelength the higher the note. Now when the car is approaching, each crest and each trough starts from slightly closer, because of the movement of the car, so that the wavelength is shortened and the note is higher than normal. When the car has passed, the reverse process takes place, and the note drops to lower than normal. It is a similar process with the emission of light. The wavelength is increased or decreased according to whether the light source and observer are approaching or receding from one another.

When space-travellers begin to go at speeds comparable with that of light itself, a star to which the probe is moving will turn bluer and bluer as the speed increases. On the other hand, light from the earth, which he is leaving, will become redder and redder. If the speed gets great enough the light from both sources will pass from the band visible to the human eye, and both will become invisible. Special instruments will have to be devized to keep the points of departure and destination in view of, and it will be a strange universe which will appear to the eye of the astronaut.

We can now turn to some of the more mundane and comprehensible aspects of light, as it affects astronomy. The first phenomenon to be noted is that of refraction. As light passes from a less dense to a more dense medium, it is bent or refracted. The principle is easily demonstrated. If one takes a tray or board and glues on to it a strip of sandpaper, cut in a diagonal, as shown in the drawing, and then tilts the tray and rolls a cyclinder down it, the cylinder will swing away in the direction of the thickening layer of sandpaper. The sandpaper is making a surface which presents more friction, and so more resistance to the rolling. One might say that the roller is passing diagonally into a denser medium. (Fig 10)

A beam of light behaves in the same way if it falls diagonally on the surface of glass. When it emerges, the opposite process takes place. By using a prism, the light can be made to refract twice. Now, the amount by which a beam is refracted depends upon its wavelength and so upon its colour. Red is refracted least, and violet most. White light, as Newton showed, will be split up into its constituent colours of the rainbow.

There is another property of light of great importance in astronomy. The atoms of any element, when in the form of a tenuous gas, emit light of quite specific wavelengths when excited. Each element emits waves of its own individual lengths, which are characteristic of that element, and of no other. Everyone is familiar with the reddish light from a neon tube, and the bluish light from mercury vapour. A piece of copper, heated in a flame, gives off a quite-distinctive greenish colour. The spectrum of an element, in these circumstances, will tell us at once what element it is. In this way, it is possible to get information about the elements present in bodies too far away from us to examine them directly.

Spectroscopy has another use. As has been mentioned, atoms as well as radiating can absorb energy, and the energy absorbed will be the wavelength of the element concerned. Light from stars often passes through a cloud of gas at its surface, or possibly through some cloud in the intervening space. In the process of passing through, energy is likely to be absorbed, and the

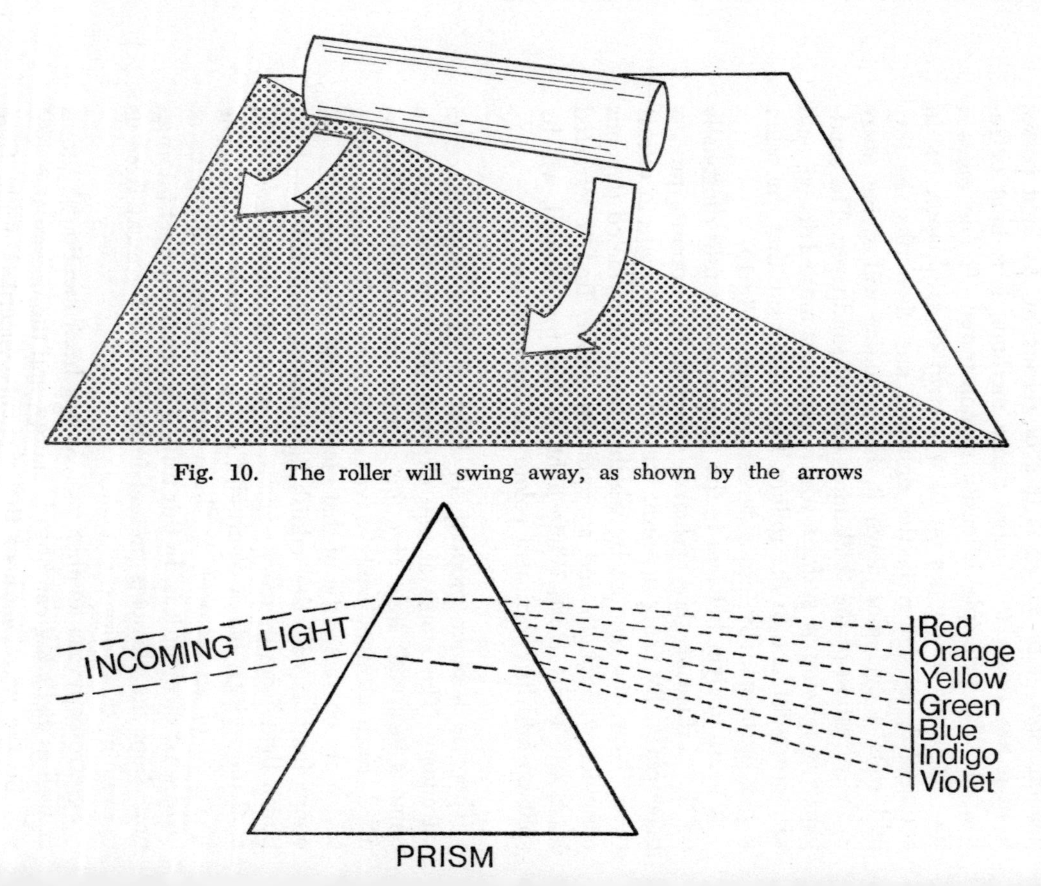

Fig. 10.   The roller will swing away, as shown by the arrows

line of colour in the spectrum which would be there if the light were uninterrupted will be replaced by a dark line. In practice, there is generally some light coming at the wavelength concerned; but it will be so reduced as to appear dark. A spectrum of a star, therefore, generally consists of a continuous band with emission lines and dark absorption lines. So we get information about both the star itself and about any intervening layers of gas.

The phenomenon of refraction has another major use in astronomy. It was by making use of refraction that the first telescopes were constructed. The largest telescopes are reflectors; but even in a reflecting telescope, refracting lenses are used in the eye-pieces. The principle of a refractor is quite simple, as the drawing shows. If we take a convex lens and allow light from some fairly distant object to fall upon it, the light coming along the middle path and striking the centre of the lens will not be refracted, because it is not meeting it at an angle, but will pass straight through. Light striking the bottom of the lens will be bent upwards, and that striking the top will be bent downwards. The three beams will be so bent that they all meet again at a point not far behind the lens. As can be seen from figure 14, the total effect is to produce a series of beams of refracted light behind the lens, with the light coming from any point on the distant object collected at another point behind the lens. In this way an image of the object, inverted, is produced. This is easily demonstrated by holding any convex lens up to the light from a window and moving a piece of translucent paper behind it, until the point is found where an inverted image of the window is formed on the paper. The more the curvature of the lens, the closer the paper will have to be held, and the smaller will be the inverted image. This image is called a real image.

If a similar lens is held close up to an object, say the lines of print in a book, the print will appear larger. In this case, the greater the curvature of the lens, the greater will be the magnification. The drawing will show why this happens. The light from the object is no longer brought to a focus behind the lens; but made to appear as though it were coming from a larger, but otherwise identical, object on the same side as the object from the lens. This is known as an aerial or virtual image.

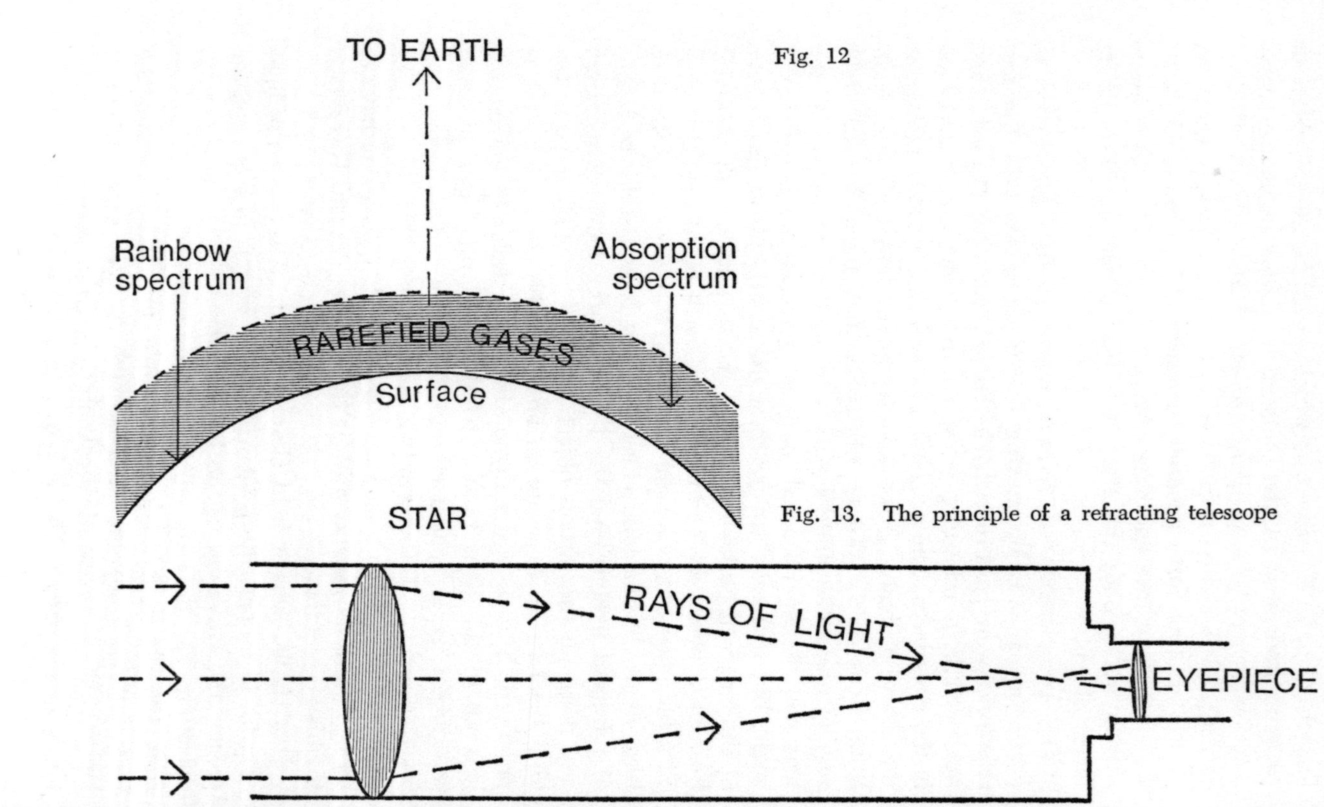

Fig. 12

Fig. 13. The principle of a refracting telescope

A concave lens has the opposite effect of producing a reduced virtual image. (Figs 14, 15)

Light rays or other forms of electro-magnetic waves have two other uses in astronomy. They are used in different ways for measuring both time and distance. For some purposes astronomy, like other branches of physics, needs fantastically accurate timing devices, and light waves provide a ready-to-hand tool. As we shall see later, the emission of light is a far more accurate measure of time than the standard from which the idea of time arises. By modern standards, the rate of the earth's revolution, on which the day, the hour, minutes and seconds are based, is not a very reliable form of clock.

When we are dealing with, for instance, the processes of some forms of radio-active disintegration, the times involved are in such small fractions of a second that a new unit is necessary in order to avoid clumsiness in specifying them. The second is now customarily divided into milliseconds, that is thousandths of a second; micro-seconds, or millionths of a second; and nano-seconds, which are thousandths of a millionth of a second. Still smaller is the pico-second or million millionth.

Whilst we are dealing with small quantities, we should also mention those of length. For some purposes, a millimetre is a gross unit, so it is divided into microns, generally written as Mu, each of which is one thousandth of a millimetre, and Ångströms, each of which is one ten-millionth of a millimetre. For the record, visible light, which is of fairly long-wavelength compared to say, gamma rays, varies between about four thousand and less than eight thousand Ångströms in wavelength. One can see that a nano-second is still quite a long time, judged by some standards, when one reckons that it takes about a millionth of a nano-second for one wave of red light to be emitted.

A large part of astronomy, as well as dealing with very short distances and time, also deals in very long ones. Our familiar world is somewhere about midway between the microcosmic and macrocosmic realms. For dealing with short distances (in the long scale), the Astronomical Unit is sometimes used. That is the average distance of the sun from the earth, or about ninety-three million miles. (92,957,209.) The next, and most usual unit is the light year, which is the distance light would

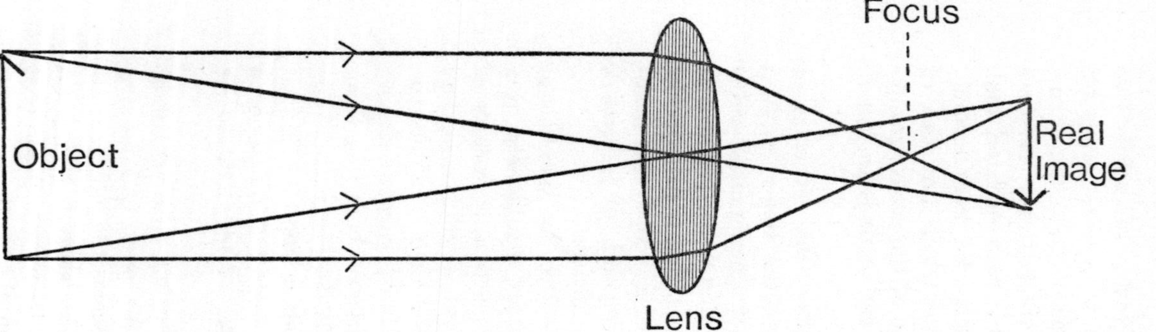

Fig. 14. A relatively distant object behind a convex lens makes a real, inverted and reduced image in front of it

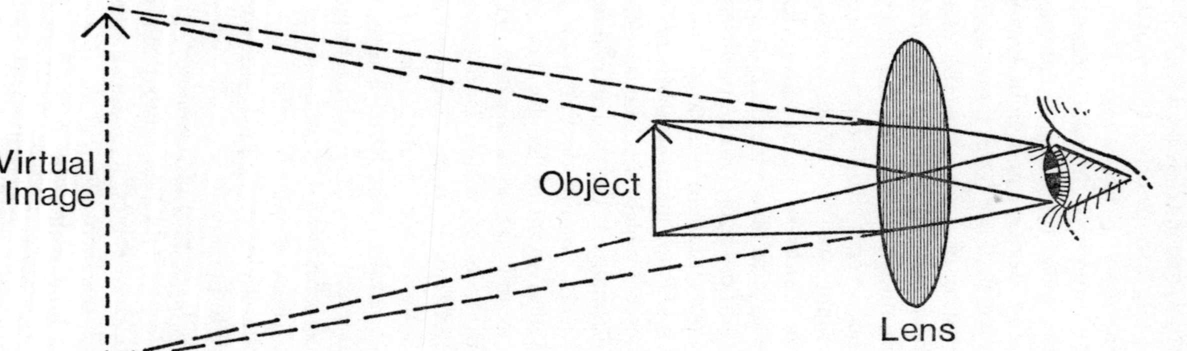

Fig. 15. An object held relatively close behind a convex lens will produce a magnified virtual image

travel in a year at three hundred thousand kilometres a second. This works out at about six million million miles (or ten million million kilometres. (9.4607 x 10^{12} Km.)

A larger unit is the parsec. This is a less straightforward unit, but it has its uses. It is the measure of distance at which a star would have one second of arc of parallax from the opposite ends of the earth's orbit round the sun, and is equivalent to about thirty million million kilometres (30.857 x 10^{12} Km.), or three and a quarter light years. Parallax is the apparent movement of a close object as seen against a more distant one, as when in a moving car the moon seems to keep pace with one, or the distant trees appear to move behind the close hedge.

Another measure used in astronomy, and some other disciplines is that of temperature. Temperature is merely a measure of the movement of molecules. In a gas, if one could see them, the molecules are all rushing about in random directions, constantly colliding with one another. The greater the average velocity, the greater the temperature. Now it is perfectly possible for the molecules to have no movement at all, though in that case the gas would have become a solid. This state represents absolute zero, which is equivalent to $-273°$ centigrade. The usual scale used in astronomy is in degrees centigrade, but commencing at absolute zero instead of at the freezing point of water and is called the K. (Kelvin) scale. Thus 0°C. is equivalent to 273°K.

When dealing with the temperatures of matter in a rarefied state, for instance during a space flight, one's ordinary ideas of temperature are misleading. The feeling of heat or cold, or, indeed, the capacity of a heat source to, say, melt ice or even metal, is a matter of the number of impinging molecules, as well as of their velocity. Thus, in an environment of very attenuated gas, the few molecules present might be moving at fantastic speeds, whilst a human being might freeze to death although the nominal temperature might be in hundreds, or even thousands, of degrees.

One other point should, perhaps, be mentioned in this short survey of the more-important aspects of physics as used in astronomy. So far the words speed and velocity have been used as almost interchangeable. In fact one should use the word velocity most of the time, though speed is the word we normally use in daily life for measuring movement. The speedometer of a car registers its speed. Velocity, on the other hand, takes account of direction as well as of speed, velocity being the speed of an object in a given direction.

Before concluding the chapter, an apology must be offered for some deliberate and important omissions. There is a great deal of electrical and magnetic theory, of great use to astronomers, which has been left out altogether. This is because a beginners' book must, of necessity, confine itself to language and concepts easily explained to a beginner. Relativity theory is basic to advanced astronomy; but it is not the sort of thing which can be explained simply or shortly. Ordinary maths and elementary physics will carry one a surprisingly long way in astronomy; but there is a point beyond which the maths and physics become almost suddenly the province of advanced professionals.

Anyone who would like to have a deeper understanding of the matters referred to in this chapter would be well advised to read Colin Ronan's admirable and lucid book *Invisible Astronomy*. He has a rare gift for making complex problems comprehensible to the layman, and deals at much greater length with a number of the subjects which have only been dealt with in outline here.

CHAPTER III

The Earth

It is one of the stranger freaks of astronomy that though we know a lot about the minute structure of matter and about the far reaches of the distant universe, we know comparatively little about our own earth. We do know its age fairly accurately from the study of the degree of disintegration of radioactive elements, and from other sources. We can be reasonably sure that it is near enough four thousand seven hundred million years old. But we do not know how it came into existence.

Two possibilities have been fancied, though the relative probability of either changes from time to time as knowledge grows and theory develops. One idea is that the earth, together with the other planets, was once a part of the sun. The sun rotates moderately rapidly, and rotated more rapidly in the past. It could have spun so fast that small chunks of matter were thrown off, rather as a speck of mud is cast off from a rotating wheel. These lumps might have condensed to form the planets. However, there are mathematical objections which have led most scientists to give up the whole theory.

The other possibility derives from the fact that space is not totally empty, even apart from the stars and planets. There are clouds of hydrogen and what is called cosmic dust. Whilst passing through such a cloud, the sun might have attracted to itself by gravitational pull quantities of hydrogen and dust, which could well have condensed into solid bodies. In the latter case, the earth would have been cold when it was formed and never have been hotter inside than it is today.

In either case the original earth, as well as the other planets, would have been much larger, and composed mainly of hydrogen. The smaller a planet is, the quicker it would lose gas of any sort. As we have noted, gas consists of molecules with random movement, and wide ranges of molecular velocities, each one of which will be affected by the results of its collisions with other molecules. By chance, and from time to time, a particular molecule may be moving so fast that it reaches escape velocity, a concept with which we will deal in greater detail later on. Any molecule reaching escape velocity will fly off until it is substantially out of the gravitational field and be lost for ever. The smaller planets have lost most of their original atmosphere, Mercury all of it. The giant planets, on the other hand, have retained a lot of atmosphere, and it may be that Saturn has no solid core at all.

The earth is certainly hot inside now. The temperature rises steadily as one delves down into it. At the centre it must be very hot indeed. If it was originally part of the sun, this is easy enough to understand. If, however, it were formed cold, we must look for another explanation. There are two candidates. Mere gravitational pressure will cause heating, and in a star the heating due to gravitational contraction is enough to set off the thermo-nuclear process.

In the second place, there is always a certain amount of heating going on inside the earth from the decay of radioactive substances.

As the earth is a massive body, very little of the heat so generated far down would escape from the surface. In consequence, there would be a process of the gradual build-up of heat. Both these processes would make the inside of the earth hot.

The earth is not quite a sphere, though it is very nearly one. All rotating bodies have a tendency to flatten out in the centre, at right angles to the axis of rotation. The diameter of the earth at the equator is very nearly four thousand miles, or six and a third thousand kilometres (6,378 kilometres). At the poles it is nearly twenty-two kilometres less. It has a mass of 5.976 x 10^{27} grammes; about one millionth of that of the sun.

Relatively speaking, we know very little about what it is like inside. What we do know, we have learned fairly lately, chiefly from observing the behaviour of seismic waves, whether natural or artificial, passing through it. The outer shell is called the crust and is only thin, varying from a mile or two thick under some of the oceans to twenty miles or so under the great land masses. Under this thin crust, there is a very thick layer known as the mantle, made mostly of rock, under which comes the core, possibly mainly of iron. The boundaries between these three shells, or layers, seem to be sharp and sudden. It may well be that the core is liquid or even possibly liquid with some solid in the centre.

Though the mantle is solid rock, it appears in some way to be partially viscous. A longish time ago there was a mad theory that the continents of the earth drifted about over the surface. As does occasionally happen in the history of science, the 'mad' idea of yesterday becomes the latest fact or hypothesis of today. So it has proved with the drifting continents. The evidence that the continents have drifted about the surface of the earth is becoming overwhelming.

The earth travels round the sun, completing one journey a year. Like all planets, it moves in an ellipse, though it is not far from a circle. At aphelion, the point where it is furthest from the sun, it is only three million miles further from the sun than at perihelion, its closest approach. Oddly, at first sight, we are closer to the sun in the northern winter than in the summer.

This is because the axis around which the earth itself revolves is inclined at 23½° to the ecliptic – the plane of the path in which the earth travels round the sun. In the winter, that is the northern winter, the north pole is inclined by this amount away from the sun. The sun, therefore, rises less high in the sky, and the heat rays from it home obliquely, and have to pass through a thicker layer of atmosphere to which they lose heat. The opposite effects are in action in the northern summer.

Unfortunately for the purpose of simplicity, the movements of the earth are not as tidy as they might be. In the first place, it does not complete a whole number of turns on its axis in one year. In consequence, the year is not a whole number of days. If we take the year as the time taken to go from one point of spring equinox to another, which is the ordinary year with which we generally deal, and which is called a tropical year, the length of the year is about 365¼ days – actually 365.24219.

The equinox is the moment when the axis of the earth's rotation is pointing in the direction of its travel round the sun.

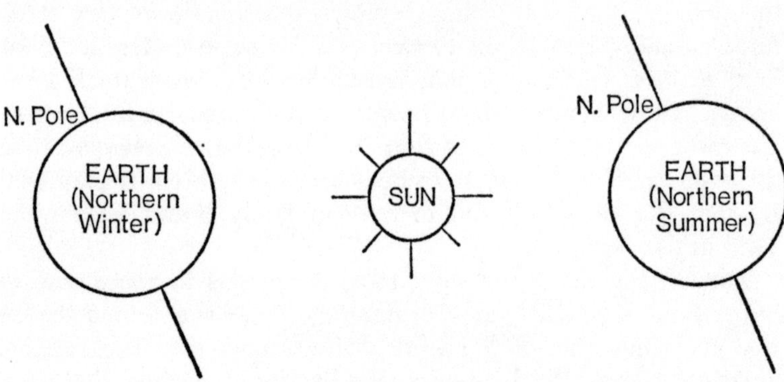

Fig. 16.   In the summer, the pole slopes towards the sun, in the winter away from it

In consequence the sunlight exactly reaches the two poles, and the days and nights are of equal length. The spring, or vernal, equinox is the time when this happens in the spring, as opposed to the autumn equinox which, as the name implies, takes place in the autumn.

Now the earth as well as spinning on its axis is also precessing, that is to say wobbling like a top before it runs down. The axis is very slowly sweeping out an arc, which takes over twenty thousand years to complete a circle. As a result, the equinox occurs at a very slightly different place in the earth's orbit every year. It also follows that one complete orbit, as viewed from a distant star, is minutely different from an orbit judged as from equinox to equinox. This year, known as a sidereal year, is equal to 365.25636 days.

That does not quite exhaust the complications. The elliptical orbit of the earth round the sun is itself slowly turning round, so that the time of one pearihelion to another, is therefore slightly different from the two other kinds of year. This year, the anomalistic year, is 365.25964 days long.

There is a last, though very minor, variation in the earth's rotation. The arc swept out by the axis in its precession is itself slightly on the wobble, so that the circle is made up of minor curves. This is known as nutation.

The matter of the day is not quite so complicated as the year; but it is not entirely simple. A civil day is the period taken, on average, for a particular spot on the earth to be exactly facing the sun; that is the time when the sun is exactly south of the spot, or in transit, until it is in transit again the next day. In fact, this average time represents rather more than one complete rotation of the earth, and is due to the fact that the earth is also moving round the sun. Perhaps the easiest way to visualise this is to suppose that one is standing in a slowly moving train, turning round on one's heels. If one starts a turn just as one is facing a telegraph pole, and according to which way one is turning and the train moving, one will take either more or less than one complete turn before facing the pole again. The directions of the earth's rotation and orbit are such that it takes rather more than one rotation before the same spot on earth is exactly facing the sun again.

Unfortunately, even this period is not regular. As we noted earlier, the rate of the earth's movement round the sun varies according to which bit of the ellipse it happens to be on — faster at pearihelion and slower at aphelion. As a result, the time taken between two transits of the sun varies with the period of the year; sometimes it is more and sometimes less than the average. As we have said, the civil day is the *average* time of a rotation judged in this way.

For purposes of astronomical observations, the civil day is not of much use, and the sidereal day is used as the basis for astronomical clocks. The sidereal day is the time in which one true rotation of the earth takes place, as seen from a distant star. This is the same thing as saying that it is the period taken for a given star to be due south, or in transit, twice, from one transit to another. The sidereal day is, obviously, slightly less than a civil day, being a fraction more than twenty-three hours fifty-six minutes. Naturally then, the stars rise just about four minutes earlier every day, as judged by civil time. Over a period of a year, each star in turn will transit at a given time, say midnight.

There is one last, if somewhat technical, consideration about time. Apart from minor irregularities, the earth is slowing down all the time, retarded by the friction of the tides. The amount of the retardation is extremely small, measured in ordinary units, and only amounts to about a hundredth of a second a century. The slowing, however, is cumulative. Over historic times, the earth has lost about one-quarter of a revolution, and this is easily detected from old records of eclipses. Each hour that passes is slightly longer than the previous one. Actually, by about ten nanoseconds – which makes it sound quite a lot.

There is one consequential problem, which need not concern us, though it presents an interesting philosophical problem. The unit of time is based upon the rotation of the earth. Using that as a clock, we have found that the emission of light waves is extremely regular, and we use such a derived standard to check the rotation of the earth – and then say that that is inaccurate! But that is just one of the paradoxes of modern science.

From the astronomer's point of view, the earth is chiefly important as the platform from which we make our observations of the universe, for which purpose it is necessary to provide the surface with a lattice of reference points. A sphere offers, *a priori*, no convenient points upon which to base such a reference system; the earth, however, since it rotates about the fixed poles, can have a system based upon these.

A series of great circles, passing through the two poles, divides the zones, rather like the liths of an orange, which are known as lines of longitude. A great circle is a circle drawn on the surface of a sphere, and which is the largest circle that can be so drawn; if the sphere were cut through along a great circle, it would be divided into two equal parts. Each great circle, therefore, marks a circumference. Longitude is measured in degrees east and west from the prime meridian. Some place had to be chosen arbitrarily, and Greenwich was chosen by international agreement as the position of what came to be recognized as the leading observatory.

In the other direction, the earth is divided into zones by degrees of latitude, numbered from a great circle, called the equator, which is at right angles to the lines of longitude, and midway between the poles. Each circle of latitude, north and south of the equator, is shorter than the last. Starting with a great circle at the equator, the length of a parallel of latitude shrinks to zero at the poles.

Each spot on the earth can be uniquely defined as to position by specifying both its exact latitude and longitude. Incidentally,

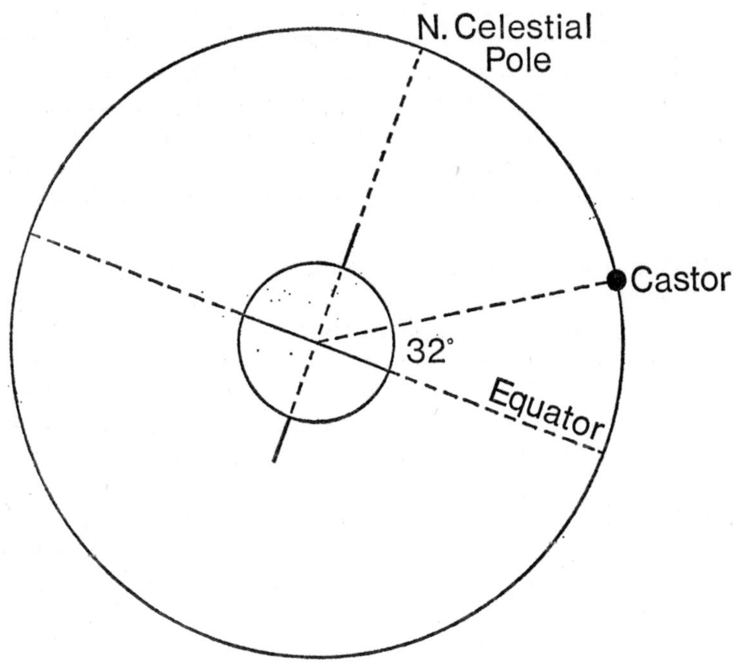

Fig. 17. The star Castor has a declination of +32, and appears 32°
north of the celestial equator

one nautical mile is equivalent to one minute of arc on a great
circle. The circumference of the earth at the equator is therefore
360 x 60 nautical miles, which is the number of minutes in a
complete circle, and amounts to 21,600 nautical miles. The nautical mile is used in navigation, and is longer than a statute
mile, having 6,080 feet compared with 5,280 feet in a statute
mile.

The celestial sphere is marked out in a precisely similar way,
so that the exact position of any heavenly body can be uniquely
specified. The position lines corresponding to terrestial latitude
are known as declination circles and could be shown as mere
projections of lines of latitude. The declination of a celestial
object is defined in degrees north or south of the celestial equator,
which itself is a projection on the celestial sphere of the terrestial equator.

The equivalent of terrestial longitude on the heavenly sphere is right ascension, which is measured in hours and minutes, instead of degrees, from an arbitrary point on the celestial sphere known as the First Point of Aries. Just as there is no natural point to choose for reference for longitude, so there was no natural point to choose on the celestial sphere as zero for right ascension. How the First Point of Aries is defined will be discussed later.

It may seem strange to be talking about the celestial sphere, as though we were some ancient people, who believed that there was truly a glass sphere surrounding the earth, and to which the stars were affixed. But, as a matter of convenience, we are quite justified in doing so; we merely use the concept as a convenient fiction, which fulfils all the needs of celestial cartography. We treat the stars as fixed on the sphere, which appears to rotate around the earth once every sidereal day. The sun, the moon and the planets apparently move about on the sphere, with constantly changing right ascensions and declinations – just as a ship, moving across an ocean, constantly changes its latitude and longitude.

The positions of the moon and the planets on the celestial sphere can be directly seen and calculated; but the presence of the sun precludes the possibility of seeing the stars at the same time. Nevertheless, the sun at any given moment is in some specific part of the heavens, and it is quite easy to calculate where, so that it can be assigned a declination and right ascension, just like any other heavenly body.

To define the First Point of Aries is slightly complicated by the phenomenon of precession. It *was* the position in right ascension of the sun at the moment of the vernal equinox, which was in the constellation of Aries, the Ram – hence the name. The whole lattice of right ascension and declination depends for its position, by definition, upon the projection of the poles of the earth on to the celestial sphere. But, as we have noted, the axis of the earth is slowly wobbling or precessing, and the whole latticework must necessarily wobble or precess with it. At the present time observers in the northern hemisphere are lucky in that the north celestial pole is within a degree of the fairly bright star, Polaris, known as the Pole Star, so that the point is easily picked out by eye. With the precession of the

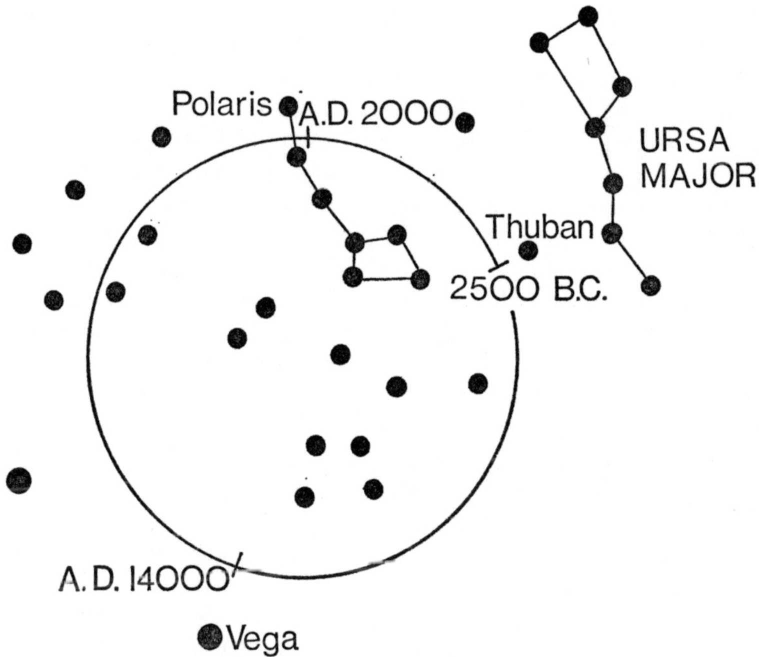

Fig. 18.   The circle shows the track of the celestial pole as it moves
slowly round, due to the precession of the earth

earth, the celestial pole will slowly move away from Polaris, and
future generations will lack this fortuitous marker.

Since, however, the whole lattice moves with precession, it
follows that the First Point of Aries is itself moving, and is
now in the constellation of Pisces. All star atlases are necessarily
dated, so that adjustments can be made for the effects of precession. Though one whole circle takes about twenty-six thousand
years, the very accurate measurements necessary in astronomy
require adjustment to be made in star maps with regularity.

There is one other long-term adjustment which needs to be
noted here. We talk about the 'fixed' stars, in contradistinction
to the moving planets, but the word 'fixed' is only relative. In
fact, and as we shall see later, the stars are moving about at
great velocities; they only seem to be stationary because they
are at such immense distances from us that their movement is
not apparent over shortish intervals of time. Naturally, the

nearer a star is to us, the sooner we shall notice its movement. Some stars then will seem to move faster than others. Over great periods of time, the shapes of the constellations will slowly change. Over shorter periods, the movement will be less apparent; but the very close stars will move appreciably. This visible movement is known as proper motion, and needs to be taken into account in fixing very exact positions for the nearer stars, though amateur astronomers need not worry their heads unduly.

One other feature of the celestial sphere needs mention: this is the ecliptic. The ecliptic is the apparent path of the sun through the star field, and is inclined at 23½° to the celestial equator. This tilt of the sun's path is caused by the revolution of the earth round the sun, its track being in a plane inclined at 23½° to the direction in which the axis of the earth is placed. Since the planets (apart from Pluto) move round the sun on roughly the same plane as the earth, they are never very far from the ecliptic as seen in the sky.

The earth rotates from west to east; naturally, therefore, the celestial sphere seems to move from east to west. To find the position of any heavenly body, one needs to know its right ascension and declination. Declination is easy, it is the number of degrees the body in question is north or south of the celestial equator. Alternatively, one can subtract the declination from 90°, and so get its distance from the pole if it is north and add 90° if it is south.

Right ascension is slightly more complicated. One starts by finding the civil time of the transit of the First Point of Aries (unless one is fortunate in having an astronomical clock, which keeps sidereal time). This is the moment when the prime celestial meridian is due south at Greenwich, and the time can be found from most almanacs, and all nautical almanacs. The right ascension of a star is the number of hours, minutes and seconds later when the star in question will be in transit at Greenwich.

If one does not live on the meridian of Greenwich, one naturally needs to add or subtract the time difference of one's local position. For every degree one lives east of Greenwich, one subtracts four minutes. For every degree one lives to the west, one adds four minutes, since that is the length of time it will take for a star in transit at Greenwich to reach one's position west of it. If the distance is very great, and one needs exact accuracy, one needs to remember that the four minutes is the fraction of the sidereal and not the civil day, the difference being just under four minutes in twenty-four hours.

The same correction needs to be made for checking time by the sun, which is one of the two reasons why sundials are not very accurate. If, for instance, one lives in Penzance, the sun will be in transit nearly twenty minutes later than it is at Greenwich; and the sundial must be set by local time and not Greenwich time. If one lives in a very large country like the United States, the differences are so large that the country has to be divided up into zones of time, each one hour later than the other. On the boundary between two zones, a sundial will be wrong by an hour, unless set to local time.

The sundial will also be inaccurate by different amounts at different seasons of the year. This is because, as we noted, earlier, the earth is moving round the earth at varying speeds. Noon at Greenwich is when the *average* sun would be due south. The real sun may be up to nearly twenty minutes out of step with the average sun. In setting a sundial, or checking local time, allowance must be made for the difference between the mean, or average, sun and the real sun. Any almanac will give the difference for each day of the year. True noon may be either earlier or later than noon as judged by when the sun is in transit.

Reference was made earlier to the effect of the tides upon the rotation of the earth. The moon, like any other substantial body, has its own gravitational field, and this affects objects on the earth. Water, being mobile, reacts to the moon's gravitational field. At first sight, it is easy to imagine that the moon's pull bunches up the water of the oceans towards itself, so causing a tide. The real picture is not quite so simple. If one searches through the books and encyclopaedias, one finds a series of explanations, most of which are hard to follow, and some of which appear to contradict others. Most of the apparent contradictions are due to the use of words and the presentations of physical laws in different forms. The full explanation is a rather complex one, and not really relevant to astronomy.

Actually the amount of the rise in the deep ocean is very small; but as the tide sweeps in over the Continental shelves,

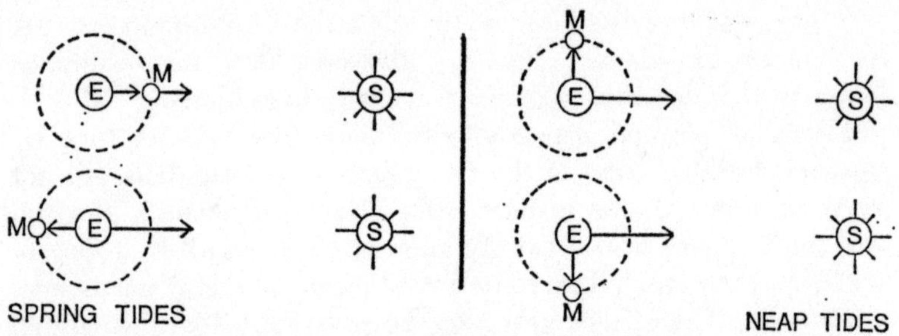

Fig. 19. At spring tides the sun and moon are pulling in the same line, and make extra high tides. At neaps, the attraction of the sun is acting against that of the moon, and the tides are therefore smaller

it piles up. In the Channel Islands, for instance, it can be as much as 40 feet, and even more in the Bay of Fundy.

The sun also attracts the earth and causes tides. Although the sun is further away from the earth, it is so much more massive that its pull is greater than that of the moon. However, what we are concerned with is not the strength of the pull, but the difference between the average pull on the centre of the earth and the greater and less-than-average pulls on the side directly facing, and directly away from, the sun. Owing to the sun's great distance from the earth, the difference is less than the difference in the moon's pull, so the sun causes smaller tides. When the sun and moon are on the same side as the earth, at new moon, or opposite to one another, at full moon, the lunar and solar tides are added to one another. We then get spring tides, that is to say exceptionally high tides, with the water coming in further and going out further. When the two bodies are at right angles to one another, at the first and last quarters of the moon, the tides tend to cancel out, and we get neap tides, that is to say exceptionally low ones.

CHAPTER IV

The Moon

The moon has always been an object of special interest to amateur astronomers, and until very recently it was hardly studied by professionals. Being so close, and appearing so large, it is easily studied with small telescopes. Moreover, the limits of observing power are largely set for the moon by atmospheric turbulence, rather than by the size of the telescope – a point we will deal with under the heading of Observing. One can see the moon through a twelve-inch reflector about as clearly as it can be photographed by the largest instrument in the world: the 200 inch at Mount Palomar.

The moon is particularly photogenic, and the vast number of photographs taken by amateurs formed the great part of the knowledge we had of what its surface is like. Unfortunately for this form of exercise, the fun has been rather spoiled by space-travel. There is no great value in even the best photographs taken from the earth, or the best recordings of observations, now that man goes to our satellite, and examines it at first hand. Moreover, these journeys, and the actual specimens of moon brought back, have unleashed a vast amount of scientific interest and experiment. In one sense, the moon has come into its own.

All this means that it is a very difficult subject to write about, when all our ideas about the moon and its structure are in the melting pot. Old controversies have flared up into immediate subjects for arguments, based on direct knowledge. It is clear, though little else is, the next few years will see our ideas about the moon clarified, and many of the differences of opinion resolved. Anything written at the moment is likely to be out of date before the book is in print.

One thing seems to be emerging pretty clearly, and that is that the age of the moon is almost exactly the same as that of the earth – about 4.7 thousand million years. We still, however, have no more clear idea of how it was formed than we have of the earth's origin. It is even remotely possible, though the idea is unlikely in the extreme, that the moon was once part of the earth. Whether, however, it was formed as part of the process of the birth of the earth; whether it was originally hot or cold; or whether it was formed quite separately from the earth, we do not know at the moment. It may well have been formed separately, as part of the birth of the solar system, and then 'captured' by the earth's gravitational field. Indeed, there are almost endless possibilities. One thing to be borne in mind is that the earth is not unique in having its own satellite; several of the planets have them. Jupiter, indeed, has no less than twelve moons of its own, though admittedly eight of them are very small.

We generally speak of the moon as being in orbit round the earth, but, as we have noted earlier, this is not strictly true. All celestial bodies which move in concert with one another move in fact round their common centre of gravity. If one thinks of spinning a dumb-bell round on a pivot, one can see that the pivot would have to be midway between the two ends. If, however, we want to balance an asymmetric dumb-bell, we shall have to choose a point nearer to the heavier end than to the lighter end. In fact, we shall be choosing the point where the common centre of gravity is fixed. With the earth-moon system, this point is known as the barycentre; the earth being so much more massive than the moon, the barycentre is actually inside the earth. As seen from space, the earth revolves, in a kind of wobble, round this point as the moon goes round it.

The actual mass of the moon is 7.351×10^{25} grammes, rather more than one-hundredth of the mass of the earth. Since the distance from the surface to the centre is very much less than that on earth, the gravitational pull at the surface of the moon is about one-sixth of that on the surface of the earth; one can jump six times as high with the same effort. The mean diameter of the moon is rather over two thousand miles, roughly a

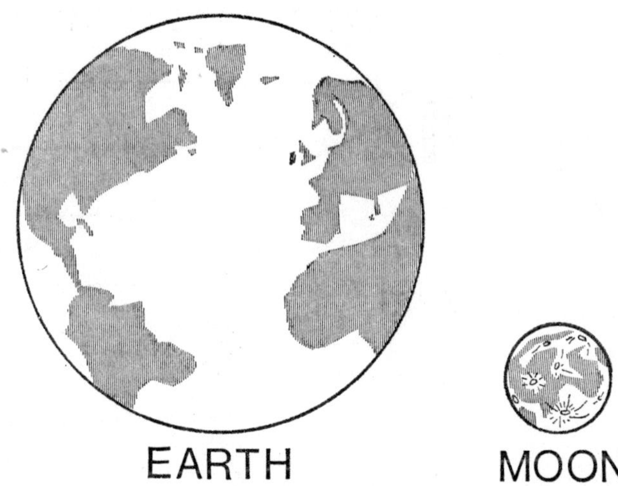

Fig. 20. The relative sizes of the earth and moon

quarter of the diameter of the earth. It can be seen that, as volume varies as the cube of the diameter, the moon has about one sixty-fourth the volume of the earth. Since the weight is proportionately less, the specific gravity of the moon must also be less than that of the earth – it is relatively lighter.

Whereas the earth is a relatively perfect sphere, the moon is slightly egg-shaped, with the bulges towards and away from the earth. This results in what is known as a captive rotation. That is to say, the moon always keeps the same face towards the earth, kept so by the pull of the earth's gravitational pull on the main bulge. Actually we see rather more than half of the moon because it moves slightly from side to side, showing us a little of what would otherwise be the hidden side. This wandering movement is called libration. One of the most exciting moments in space history was when a Russian satellite first passed behind the moon, and sent back pictures of the hidden side, which no man had ever seen before. It turned out to be not so very different from the side which always faces us.

The moon, like all heavenly bodies, moves in an ellipse, so that it is sometimes nearer to us than at other times. The difference, however, is too slight for the apparent size of the moon to be noticeable to the naked eye. The average distance is about 240,000 miles – actually 238,855 miles or 384,400 kilo-

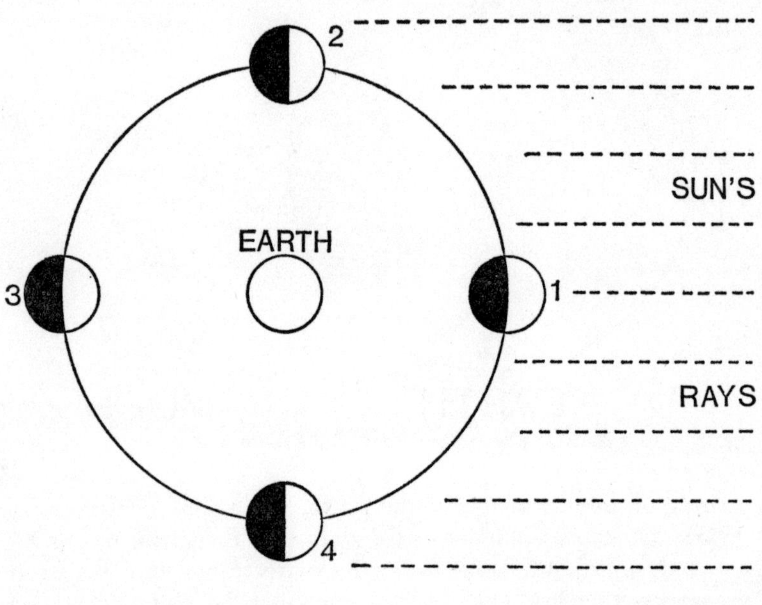

Fig. 21. Why the moon shows phases. Half of it is always illuminated; but we only see that part of the illuminated half which can be seen from the earth. At position 1 we can see nothing. At 3 the full moon is visible. At 2 and 4 we see half the disc

metres. The distance originally took a good deal of trouble to measure. Now, however, we can do it very accurately by sending radio signals and measuring how long it takes to receive the echo; the measurement, in fact, is accurate to a matter of five inches!

To make one revolution of the earth, the moon takes just over twenty-seven days: 27.32166 to be precise; but that is its sidereal period – the time for one revolution, as judged from a point in far space. Since the earth is moving round the sun, the apparent time for one revolution, that is from the moment when the earth, moon and sun are in line at new moon to the moment they are back in the same line, is quite a lot longer than the sidereal period; about twenty-nine and a half days (29.53059).

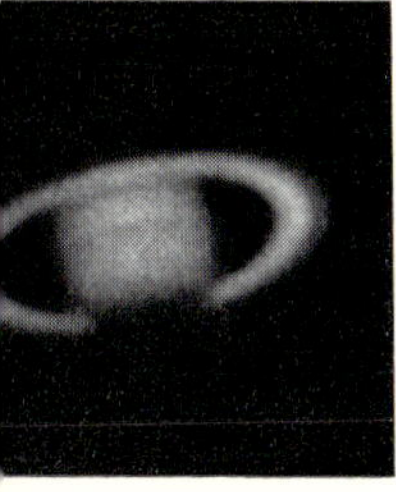

Saturn. A photograph by H. E. Dall using a 15″ telescope

Venus, photographed by the author. Note the blunted cusp cap

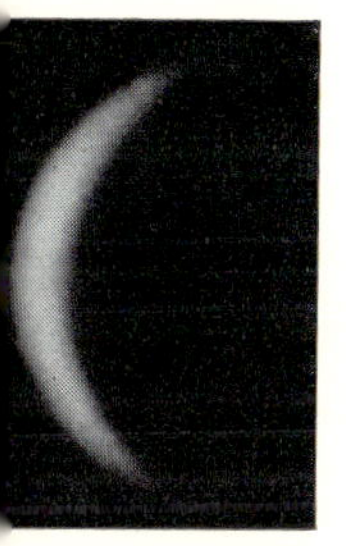

Venus, photographed by the author, when very close to the sun

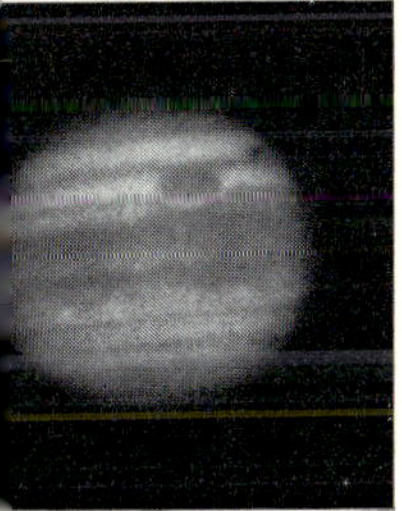

Jupiter photographed by H. E. Dall. The Great Red Spot is the oval towards the top right hand side. Further right, and just above, is the dark shadow of a satellite

The full moon showing rays. The large ray-system comes from the crater Tycho

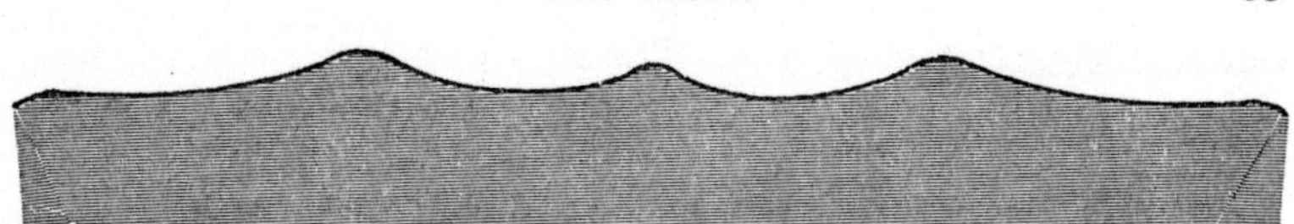

Fig. 22.   Profile of typical lunar crater. The slopes are much more
gradual than they appear when cross-lighted

This is called the synodic period, and is a lunar month, from
new moon to new moon.

The phases of the moon are simple to understand. When the
moment of the new moon comes, the moon is in direct line
towards the sun, so that the side of it which is illuminated is
directly away from the earth and is invisible from the earth. At
full moon, the earth is between the two bodies, and the whole
of the illuminated half is facing the earth. Half-way between
these two stages is the period of the half moon, when the line
moon, earth, sun is a right angle. Naturally, the full moon is
always in transit at any given spot at about midnight.

If the moon went round the earth in exactly the same plane
as the ecliptic, manifestly the moon would eclipse the sun every
time it was new; that is directly in line between the earth and
the sun. Equally, the earth would eclipse the moon at every
full moon. That this does not happen is because the planes
of the earth's orbit and the moon's orbit are inclined to one
another. An eclipse takes place only when the two cross one
another, and the moon happens to be at or very near one of
the two points of intersection. The movements of earth and
sun being regular, eclipses occur at regular intervals, which
can be calculated very exactly. The whole series of eclipses go
through a regular cycle of nineteen years.

The moon, being a small and light body, as astronomical
bodies go, has lost all of its atmosphere. If there is any remaining
at all, it is so tenuous that it can be ignored for almost all
practical purposes. This in a way is odd, because the moon
looks exactly as though its surface had been heavily weathered.
Without even a small telescope, and as seen with the naked
eye by the ancients, it looks rather like the surface of the earth.
It seems as though it were covered by continents and seas.
When seen through a telescope it looks, at first glance, very
similar, except that the 'continents' are pockmarked with in-

C

numerable craters of all sizes. The smoother areas, called maria, or seas, seem to be almost flat and smooth, though there are a few small craters even in the maria.

The 'continents' are higher than the marias, and contain numerous tall mountain ranges. Some of these rise to peaks, comparable in height to terrestial mountains. As the moon is so much smaller, the mountains are relatively taller.

One of the fiercest controversies about the moon is how the craters came to be formed. One school of thought maintains that they were formed by the impacts of meteorites. There must, in fact, be many meteoric craters on the moon since we cannot doubt that, like the earth, it is under constant bombardment. We do not see any except the very largest and newest craters on earth, because it is only at very long intervals that we are hit by giant meteorites. Even where there is a large crater formed, the wind and rain gradually erodes the crater away, and it gets filled with blown dust and particles.

Since there is no atmosphere on the moon, there can be no wind or rain to do the weathering, and a crater, once made, can endure almost indefinitely. The strange thing is that, if one looks at some of the craters through a telescope, they appear exactly as though they had been weathered. Some are so indistinct that they can hardly be distinguished. Others seem to be filled up, and others still have more recent craters superimposed on them.

The alternative theory is that the craters were formed by some sort of volcanic action. That is not to say that they are simply the craters of volcanoes; but that some kind of lava-like action has been at work. On this theory, the lava may have flowed over the maria, and formed the smooth surfaces. Through a telescope one can see that there are, in places, ripples on the surface, as might be caused by the flow hardening.

One very strong argument for the volcanic theory is provided by a series of long chains of very small craters, which seem to run along faults in the surface. It is impossible to believe that these crater chains could be caused by meteorites. Even the meteoric protagonists will admit that these features must be volcanic in origin. Another argument rests on the crater patterns. A series of observations through a telescope will demonstrate even to a beginner that the craters follow zones or patterns,

and it is hard to visualize any factor which could cause meteorites to fall in so systematic a manner.

There are many other features of the moon worth studying
through a telescope, and whose origins are still a matter of
dispute. When the moon is full, one of the most fascinating
sights is the pattern of bright rays, which are centred on a few
craters, and whose arms spread out over long distances. The
most brilliant display is the ray system from the crater Tycho.
The rays of bright material spread out over huge lunar distances,
and pass across the surfaces of many craters. Clearly, however
they were caused, they were ejected after the craters over which
they pass were formed. The rays could be some sort of volcanic
ash, thrown out by volcanic explosions; or, just possibly, splashed
out by the impact of a huge meteorite.

Other features worth looking at are the rills and valleys, some
of them looking exactly as though they were the beds of dried-
up rivers. A particularly interesting one is Schröter's Valley, a
long serpentine valley near the crater Herodotus. It resembles
a cobra, or can be seen as a river, winding, and flowing into
a small lake. As there was never any water on the moon, to
the best of our belief, the appearance is a mere coincidence.

Amongst the rather curious features are a number of 'domes'
which are slight swellings on the surface, quite smooth and
round, and looking rather like small blisters. These could well
be accounted for on the volcanic theory as bubbles of gas or
lava, solidified before they had burst. One can see how this
could happen by looking at the surface of a pan of boiling
porridge, where bubbles form, burst, and leave momentary
craters behind. Indeed, one can make a fair imitation of the
moon's surface by photographing boiling porridge in a strong
cross light.

The lunar craters, it should be noted, are far shallower than
they appear. One views them through a telescope when the
sunlight is slanting across, throwing strong shadows and setting
the walls in sharp relief. When looked at under full sunlight,
at the time of the full moon, they appear as mere shadowy
outlines. A crater is more like a shallow saucer than the kind
of crater on a terrestial volcano with its steep sides. Many of
them have central peaks, and one of these, in the crater Alphon-
sus, has an interesting and significant history. In 1958 the Russian

Kozyrev observed some kind of disturbance in this crater, and managed to take a spectrograph which is thought to have demonstrated the presence of hot gas, as though from some kind of lunar eruption. There is some dispute about the evidence; but other observers have, at intervals, reported obscurations which could be accounted for in a similar way.

Though the moon has become a suburb of the earth, and subject to direct observation, there are still vast stretches where it will be a long time before man sets foot, and which still provide the amateur with objects of study and speculation. One such is the great scar in the lunar Alps, looking as though a cannon-ball had been fired through the peaks, or as if a huge meteorite had ploughed obliquely through the range, though, in fact, it is certainly a collapse feature.

Another visually interesting feature is the so-called Straight Wall in the Mare Nubium. It is visible through quite a small telescope, and at first suggests an artifact. Closer observation shows that it is neither straight nor a wall. The irregularity of shape is small, but quite distinct, while shadow measurements prove that it is a gentle slope rather than a precipice. The slope angle is no more than about 40°.

One sight which the moon can provide is very transient, and one is lucky to catch the illumination at exactly the right moment. This is when the rising sun just catches the peaks of the Sinus Iridum, an arced bay to the north of the Mare Imbrium. Caught at the right moment, the sunlight on the peaks shows up as a flaming crescent, projecting from the limb, and can be seen clearly even through a moderate pair of binoculars.

We have spoken about the exploration of the moon by visiting astronauts; but there are so many limitations and difficulties that it is likely to be a long time before all the parts we can study through telescopes have been visited, so that there is still scope for observation from the earth. One of the difficulties facing future explorers is the result of the moon's lack of atmosphere, and the length of the lunar 'day'.

Since the period of the moon's rotation is the same as the lunar month, it follows that any part of the lunar surface is exposed to the sun for nearly fifteen days, and is then in darkness for a further fifteen days. Since there is no air to protect the surface from the heat of the sun, nor to keep in some of the heat during the long lunar night, the midday temperature at the equator is torridly hot, and the extended nights bitterly cold. Quite apart from the need for astronauts to take their own supply of air and food, the climate of the moon is thoroughly hostile.

Though by the beginning of the next century the moon may have lost some of its purpose as a study for amateur astronomers, it will provide both interest and entertainment for a long time to come. When one first buys a telescope and sets up as an observer, the moon is still the prime object of first concern. It will always remain as an object of awe and fascination to the beginner taking his first look through a telescope.

Of course, the moon is unique in that it is the only world to have been reached – so far. The first astronauts, Neil Armstrong and Edwin Aldrin, stepped out on to the surface from the lunar module of Apollo 11 on 21st July 1969. Years of observation as well as engineering research had led up to that great moment. There had been a theory that the moon's dark 'seas' were made up of soft dust, into which a space-craft would sink; the idea never really fitted the facts, and had been disproved by earlier landings with unmanned probes, but there had been no absolute guarantee that the surface was really firm everywhere. Had the astronauts' vehicle landed at a tilt, or had one of its legs sunk down into soft material, the results would have been tragic. Fortunately, nothing of the sort happened. A second landing, made in November 1969, was equally successful.

In addition to leaving scientific equipment on the moon (where it remains, and will remain indefinitely), the astronauts brought back samples of moon rock and dust for analysis. It has been found that the rock resembles basalt more closely than anything else, and that there are glassy particles in it. Seismometers, or 'moonquake recorders', seem to indicate that tremors occur in the moon's crust, though at the time of writing the evidence is somewhat conflicting. As expected, there were no signs of past life in the lunar samples; and neither is it now thought possible that the 'seas' ever contained water.

Practical lunar research is in its early stages as yet, but already it has told us a great deal about what the moon world is like. Hostile it may be; yet it is none the less fascinating for that, and to us on earth it will always have a special importance.

CHAPTER V

The Planets

We are in the same difficulty with the planets as we were with the earth and moon; we do not know how they came to be formed. Sir James Jeans put forward an attractive theory that a passing star, on a close approach to the sun, drew out a long filament of gas, which condensed into the planets. This theory, however, has had to be abandoned. Perhaps the most favoured idea, as we suggested, in connection with the formation of the earth, is that the sun, passing through a gas cloud, gathered up matter, which condensed into the whole planetary system.

Only some of the planets were known to the ancients. Planets look to the naked eye very like stars, except that they wander about the celestial sphere, whilst the stars stay apparently fixed in position. Indeed, as we noted, the name 'planets' is derived from the Greek word meaning wanderers.

All the planets, except Pluto, revolve round the sun in more or less the same plane, but at very different distances from it. Two of the planets, Mercury and Venus, have orbits inside our own and so show phases like the moon. The diagram will make the reason plain. When one of the two inside planets, known as the inferior planets, passes to the opposite side of the sun, the whole of the illuminated half is turned towards us, and the disc therefore shows the whole of itself. One says 'shows', but it needs a good telescope, well directed, to see either of them at this stage, which is known as superior conjunction. The sun is in an almost direct line between us and the planet, making it invisible to the naked eye. Naturally, too, this is the moment when the planet is furthest from us, and makes it appear at its smallest.

70

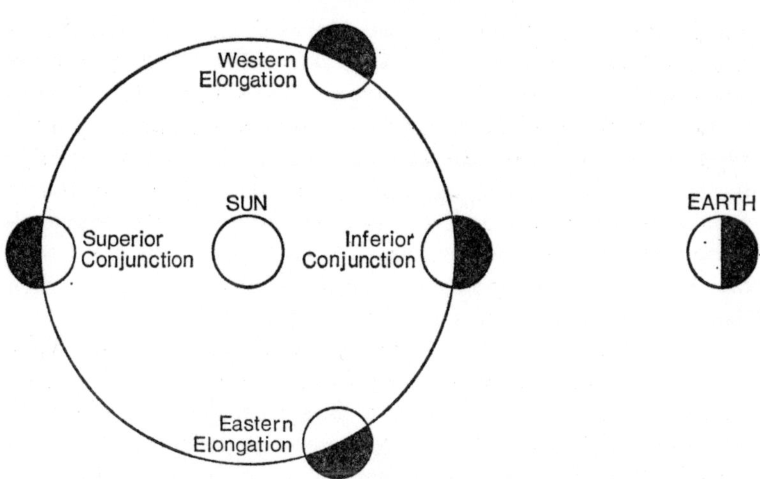

Fig. 23. The phases of Venus. At inferior conjunction the dark side is towards the earth, and we cannot see the planet. At superior conjunction the bright side is towards us, and we see the whole disc. At elongation Venus appears as a half-disc

At the half-way stage, when the earth-sun-planet lines form a right angle, the planet is at dichotomy, that is to say in a phase corresponding to the half moon. It then moves on until it is directly between us and the sun, at inferior conjunction, when none of the disc is visible. Because the plane of orbit is not quite the same as that of the earth, it is only occasionally that the lines are quite straight, and that the planet disappears behind the sun, or passes across the sun's disc at inferior conjunction; but these phenomena do occur at long intervals. Mercury is due to pass across the sun's disc in the year of writing; (1970), but will not do so again until 1973, and again in 1986.

Of the outer planets, Mars, Jupiter and Saturn were known to the ancients and all of them can easily be seen at appropriate times with the naked eye. The next planet, moving outward from the sun, is Uranus. Though it is a large planet, it is so far away that, even with a telescope of moderate power, the disc is quite small. The next planet outwards is Neptune, whose discovery makes an interesting story. Since all the planets have masses, they attract one another and make slight alterations, or perturbations, in the orbits of the other planets. When Uranus, which itself was only found by Herschel in 1781, was being examined and its orbit plotted, it was noticed that it did not move quite as it should have done: there were slight perturbations in its orbit. These could only be accounted for by the supposition that there was another quite massive planet outside its orbit. Two astronomers worked out where this new planet should be looked for and, in 1846, Neptune was discovered almost exactly in the spot where it was predicted that it would be. A real triumph for astronomical calculation in days when there were no computers to do the hard work.

In due course, the story was repeated. Both Uranus and Neptune showed further perturbations, which indicated yet another planet, still further out. It took a long time to find it. It was at last discovered from Lowell Observatory in 1930, and has been named Pluto. Unfortunately, though it was found in the right place, Pluto seems to be the wrong size. It appears to be no larger than Mars, and, unless it is incredibly dense, it will not account for the observed perturbation of the orbit of Neptune. This remains a mystery, which may be solved by the discovery of yet another planet, though it seems unlikely.

In relation to this mystery, there is another – though this could be a matter of pure chance. There is a mathematical relationship between the masses of the planets, and their distances from the sun, which is known as Bode's Law. The law has no theoretical basis; but is in good agreement with the observed facts – with two exceptions. The first exception is that there should be another planet between the orbits of Mars and Jupiter. Now the odd thing is that just about this zone is occupied by a series of small planets, sometimes called asteroids. It is not impossible that these small bodies represent the remains of a once-existing planet which was fragmented by some astronomical disaster. At any rate, the facts make the otherwise unjustified Bode's Law a matter of more than curiosity.

The other discrepancy of observed facts with Bode's Law relates to Neptune, for which the Law breaks down. We have just seen that there is a more-distant planet, Pluto, but it does not satisfy the requirements of theory, and it conflicts with Bode's Law. It is at least strange that the two conflicts should go together. If it is pure coincidence, one can only say that it is a very odd coincidence.

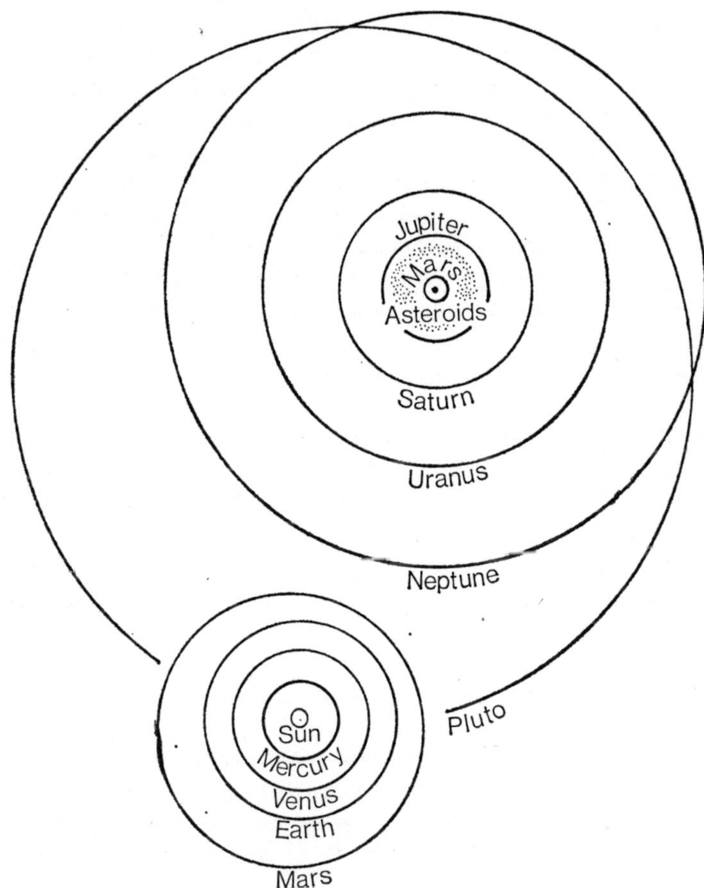

Fig. 24. The orbits of the planets. The closer ones have to be shown on a separate scale, or the paper would have to be immense

The easiest way to show the relative positions of the planets would be with a diagram. Unfortunately, the scale presents difficulties. Mercury is so relatively close to the sun, and Pluto so relatively far away, that a scale drawing showing both would be immense. The only way out of the difficulty is to show the chart in two sections: one up to and including Mars, and the other outwards from Mars to Pluto. It will be seen that whereas most of the planets have almost concentric orbits, Pluto, determined to be different, overlaps the orbits of Neptune on one side, whilst going far outwards on the other. It should, incidentally, be noted that the actual orbits are ellipses, but are shown as circles, representing the average distances from the sun.

One difference between the inferior and the outer planets is that the outer planets, except partially for Mars, do not show appreciable phases. Whether they are on the same side of the sun from the earth or on the opposite side, almost the whole of the illuminated disc can be seen. At elongation, that is when they are at right angles, there is a theoretical small phase effect. In the case of Mars, which is relatively close to us, this is easily appreciable.

Mercury, the planet closest to the sun, is an elusive object to study. In virtue of its position, it is never, as seen from the earth, very far from the sun, and can only be seen with the naked eye at sunrise or sunset when at or near elongation. Even with a telescope of moderate power, it is difficult to pick up at other times, and then only when the sky is very clear, and the telescope very well directed. As a photographic object it is very hard to capture.

One word of warning should be given to beginners. *On no account should they make efforts to find Mercury by sweeping the sky with their telescopes. Since it is always close to the sun, it would be possible to turn the telescope accidentally on to the sun itself. Even the briefest glimpse through a small telescope will result in blindness, and it is absurd to take such a risk.*

In any case, Mercury, apart from the difficulty of finding it, is of very little interest to amateurs. It is a small world, not much bigger than the moon. Being so close to the sun, it must be extremely hot on the day side. As it has a long 'day', it must also be cold on the night side. Being so small, it can have no atmosphere. Through very large telescopes, it is shown to have dim markings; but, through the average amateur telescope, all that can be seen is the phase – and that is only discernible when it is near to crescent shape. Almost certainly it will have craters, not dissimilar from those on the moon. With its extremes of temperature, and the absence of atmosphere or water, we can, at least, be sure that there is no life on this planet and that it will be an almost impossibly hostile place to visit with manned probes.

The next planet, moving outward from the sun, is in some ways one of the most interesting to amateurs. Venus, as an object for observation, has the great attraction to the lazy and those who like to be comfortable, that it is best viewed during broad daylight. Being fairly close to the sun, it is seldom visible much after sunset or much before sunrise, and is then low in the sky.

Venus, at its brightest, is much the most luminous of all the stars and planets; it is so bright that it can cast a shadow and has led to many reports of 'flying saucers' and other strange phenomena. Those with very acute vision, and who know where to look, can see Venus on a clear day with the naked eye. Even when it is very close indeed to the sun it is still possible to see it, and even photograph it, through a telescope. I have myself taken a good picture when it was within two degrees of the sun—so close that it was impossible to use the sighting telescope (see page 131) without being blinded. The same warnin needs to be emphasized here as in relation to Mercury, that it is quite unsafe to look for Venus by sweeping with a telescope. One needs a telescope with well-graded circles, (see page 132) to find it at midday.

The brightest time is when Venus is getting on for a narrow crescent. This is because, as a moment's consideration will show, the planet is closest to the earth and appears largest when it is at inferior conjunction. At superior conjunction, the full disc can be seen; but, being on the far side of the sun, it is at its greatest distance and appears quite small. There is a minor mystery connected with the dichotomy of the planet. The exact moment when this should occur can easily be calculated; but observations do not exactly coincide with theory. Dichotomy comes either a few days late or early, according to whether the disc is waxing or waning. It is probably only an optical effect; but is still interesting.

Another interesting problem about Venus is the alleged occasional appearance of an ashen light, similar to that on the moon. In the case of the moon, the matter is quite simple. The appearance of the 'old moon in the arms of the new' is caused by reflected light from the earth—it is, in fact, earthshine. In the case of Venus this is not a possible explanation; the planet is far too far away. The ashen light, unless it is a mere optical illusion, is at present unaccounted for.

In any case, the visible part of the planet exhibits other effects which have not been satisfactorily accounted for. The arms of the crescent sometimes show extensions, or cusp caps. I have myself taken a photograph which shows such a cap quite clearly, and they are often visible through telescopes of very moderate size. Some acute observers can sometimes see faint markings on the otherwise brilliant white of the visible surface. Photographs taken in ultra-violet light also show dusky markings quite sharply.

We know very little about the surface of Venus, and what we do know is from sometimes contradictory observations with reflected radio-waves, and the information sent back by probes. The whole of the surface is covered with dense white clouds, which prevent any direct observation of any markings or features which may be there. One certain thing is that the planet is very hot, since it is not only relatively close to the sun, but the clouds must act much in the fashion of a greenhouse. Life on Venus is very problematical indeed; the clouds contain large amounts of carbon dioxide, which is useless for breathing, and the immensely hot surface, probably several hundred degrees centigrade, is not inviting.

Venus has some other oddities; it revolves in the opposite direction to the earth, that is to say from east to west, which is known as retrograde motion. It proved very difficult to discover the period of its rotation because there were no surface features by which to time it. The first estimates proved to be self-contradictory; but it now appears certain that its 'day' is a very long one of 243 days; actually longer than its year of 224.7 days, which is the time for it to make one orbit round the sun.

It will save time to explain here that, when we speak of a planet's day or year, the times are given here as the equivalent length of earth days – though it sounds odd to say that Venus's day is over two-hundred days! So long as we had to study Venus from afar, our knowledge of it was bound to remain incomplete. Even as lately as 1962 it was still thought that there might be oceans there – and the possibility of life was not entirely ruled out. We know better now. Probes have bypassed the planet, and several (all Russian) have landed there, parachuting down through the dense atmosphere and continuing to transmit messages until within striking distance of the surface. The results have been discouraging. Not only is the surface temperature very high, but the atmospheric pressure is very great – at the surface, perhaps 120 times as great as the pressure of the earth's air at sea-level. And since the atmosphere of Venus is made up almost entirely of carbon dioxide, we must reluctantly cast aside any hopes of finding life there. Whether manned landings will be made in the foreseeable future is distinctly dubious. In any case, Venus seems to have been tacitly dropped from both the American and Russian space programmes so far as manned vehicles are concerned.

Outside of Venus, and beyond the earth, comes what in its way is the most interesting of the planets – the red world of Mars. At its brightest Mars is very bright indeed, though at its more distant periods it is not very remarkable. The real interest of Mars is that it is the only one of all the family of planets which is a serious candidate for the possible existence of life. In honesty, one must admit that the possibility is fairly remote, and, if there be life there, it is almost certainly a very low form of life. Perhaps bacteria, or a lesser plant life, of the order of moss.

Only when it is in fairly close approach to the earth is it easy to see markings on the surface through a moderately powerful telescope. The relative orbits of Mars and the earth result in their fairly close approach every two years, though very close approach only takes place at fairly long intervals. There will be one such good opportunity for viewing in 1971.

The dark markings on the planet are, at favourable opportunities, quite clearly recognisable, and there are good maps. In Martian winter there are white caps at the two poles, which fade in spring. They look like thin sheets of ice, or deposits of frost, which melt in the warmer weather, though in fact they may be composed of solid carbon dioxide. Looked at in a not-too-powerful telescope, there are straight lines to be seen, and 60 years ago the distinguished astronomer, Lowell, believed these to be artificially constructed canals. He even set up an observatory to study these artifacts which, to him, demonstrated the presence of intelligent beings.

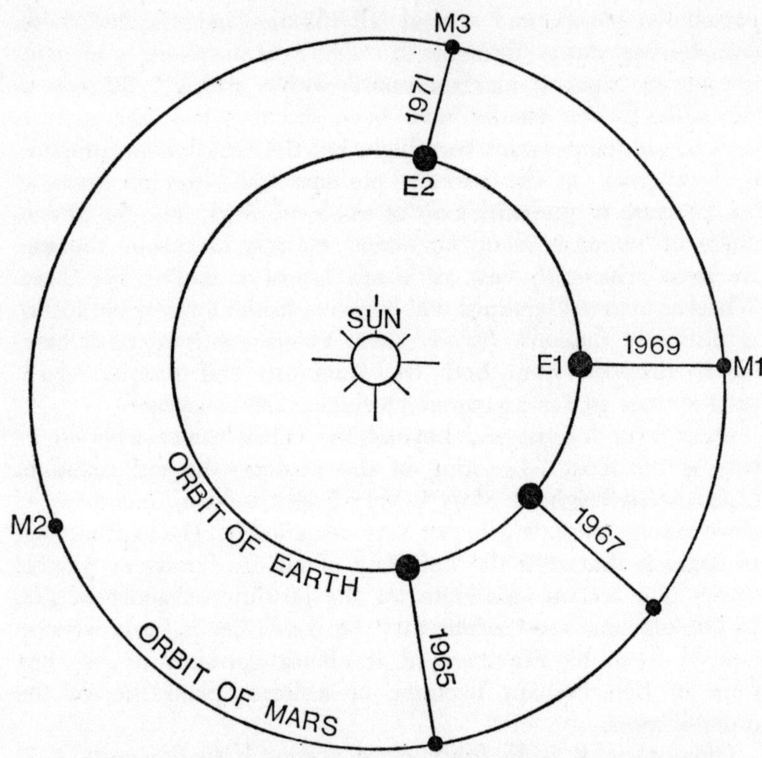

Fig. 25. Oppositions of Mars. The orbit of Mars is less circular than that of the earth. Oppositions occur when the sun, the earth and Mars are lined up; this occurred in May 1969, when the earth was at E1 and Mars at M1. One year later, the earth had completed a revolution and was back at E1; however, Mars, moving more slowly in a larger orbit, had reached only as far as M2. The next opposition, that of August 1971, takes place with the earth at E2 and Mars at M3. As Mars is then at perihelion, the opposition is very favourable (distance 35,000,000 miles), whereas at an unfavourable opposition, such as that of 1965, the opposition distance is as much as 63,000,000 miles

Unfortunately, a close study shows that these 'canals' are neither straight nor continuous, and are certainly not canals. It is a pity, because it was Lowell's canals which fed the widespread belief that Mars was inhabited. The evidence for some form of life is much more tenuous. When the polar caps melt in the spring, the dark markings on the surface tend to become darker, a change which could be due to plant life which flourishes in the warmer weather. There are, however, quite plausible alternative explanations of these changes, and the question of whether there is any form of life will have to be settled when probes have been landed and sent back detailed information.

The American Mariner probes have already passed by the planet and relayed some information and some photographs. These show that the surface of Mars is not wholly dissimilar to that of the moon. There are certainly a whole range of craters, and it begins to look as though Mars is not an attractive kind of place. In any case, it is a smallish planet, and has only a very tenuous atmosphere, and that not a very suitable one for the prospect of life. It is, of course, just possible that, in the long past, when there was more atmosphere remaining, there was life, which may now be extinct.

Whatever view one takes, and even allowing for the unlikelihood of life there, Mars must remain an object of special interest to astronomers, and particularly to amateurs. It is the one planet with surface markings clearly visible, and easily studied with modest equipment. It is, moreover, an early candidate for exploration, perhaps in the next decade or two.

Outside the orbit of Mars comes the belt of the asteriods. As we have said, these small objects could be the remains of a broken-up planet. Rather less probably they could be accretions which have failed to join up and make a planet. The largest of the asteroids is a miniature world, less than five hundred miles in diameter. The smallest we do not know, because it may be no more than pebble size or less; probably there are at least forty thousand of these minor planets. Discovering new ones is still a practical proposition.

Being so minute in size, not even the largest will have any significant gravitational field, so that there can be no atmospheres. Probably the smaller ones are not even of regular shapes; they may be mere lumps of rock or other matter.

The asteroids have, many of them, highly elliptical orbits, and some of them, upon rare occasions, come quite close to the earth – by astronomical standards, very close indeed. From time to time there are scares that we might be hit by one, though this is unlikely in the extreme. The nearest miss to date was in 1937, when Hermes a mere mile across, came closer than half a million miles. In 1968 Icarus, which is a few miles across, came within four million miles, which is still close, and in 1975 Eros will pay us a passing visit at about fifteen million miles. The only asteroid large enough, and coming close enough, to be seen with the naked eye is Vesta.

Passing beyond the tracks of the asteroids, we come to the giant planet, Jupiter, another planet of great interest to amateur astronomers. At its brightest, it is a very bright object in the sky, simply because it is so very large. It could, in fact, accommodate more than thirteen hundred bodies the size of the earth. Being so large, it has a powerful gravitational field, with an escape velocity of thirty-seven miles a second, which compares with seven for the earth. This high escape velocity means that it will have held on to all its atmosphere, which is largely hydrogen, and which is the chief constituent of the planet.

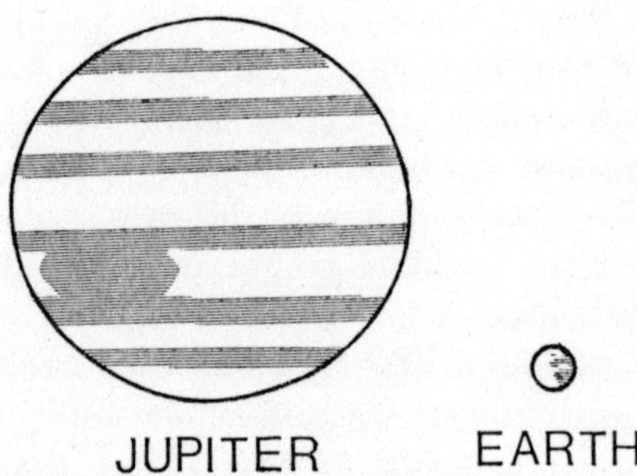

Fig. 26.   Relative sizes of earth and Jupiter

Being so gaseous, it is doubtful whether Jupiter has any solid core, and is hardly a candidate for visitors from the earth. Even if they got there, and could survive local conditions, there would be no surface to land on, and the high escape velocity would make it practically impossible to get away again. The best we can hope for is to explore it with probes, and, except for the outer layers, even this is not very practicable.

What Jupiter lacks in hospitality, it makes up for in other forms of interest. The disc itself, seen through even a small magnification, is visibly flattened. The equator is four thousand miles in diameter larger than at the poles – equal to two diameters of the moon. It possesses the grand total of twelve

moons or satellites. Four of them are so large that they can
be seen with powerful binoculars, the other eight are so small
that only a powerful telescope will reveal them. As viewing
objects, the satellites are quite fascinating. Over a period of an
hour or two, one can see them changing position, and disappearing as they pass behind or across the face of the planet.
When passing on the near side, the shadow of the satellite
can often be seen against the bright disc.

Jupiter itself rotates very rapidly, with a day of only about
ten hours. It is this rapid rotation which causes the equator to
bulge out as a result of the centrifugal force. Not being a solid
body, the different parts rotate at slightly different speeds, the
central region taking about five minutes longer than the rest.

The chief interest of Jupiter, as an object of telescopic observation, is that it is covered with cloud belts, running in
rings, parallel to the equator. The belts consist of quite sharply
marked shapes or patterns, which mostly change slowly, so that
the changes can be followed by eye. Perhaps the most interesting feature of all is the so-called 'great red spot', an oval-shaped
object of distinctive colouring, though not very noticably red.
This feature has been observed for several centuries, and
never explained. Sometimes it vanishes altogether for a few
years, but always comes back again. The cloud belts themselves
are probably made up of droplets of ammonia; but the great
red spot is too persistent in shape to be wholly liquid or gaseous.
It may be that it is some sort of very light solid body, floating
in the heavy atmosphere.

One other remarkable attribute of this planet is that, a few
years ago, it was discovered to be emitting radio waves. This
phenomenon has not, so far, been satisfactorily explained; but
it may be connected with the fact that Jupiter has, in all probability, a strong magnetic field. We shall hope to know more
about this when a probe can be sent in a few years' time.

Passing outward from Jupiter, we come to another planet of
some considerable interest to observers: the planet Saturn. In
one way, it is quite the most exciting object for a first view
through a telescope. At first glance, and if a beginner is not
told what he is looking at, it appears to be egg-shaped, with
two dark spots. When the eye is adjusted, it is seen to be a
normal spherical planet, surrounded by bright rings.

As the relative positions of the earth and the planet alter, the rings can be seen at different angles. At some periods they are edge on to the earth, and are substantially invisible. They are, in fact, only a few miles thick, though they extend outward over some thousands of miles. They appear to consist of a vast number of small objects, perhaps about the size of golf balls, in rapid rotation around the planet. Their origin is not known, but it is not impossible that they are broken fragments of a shattered satellite.

When the rings are broad on to the earth, it can be seen that they are divided into two sharp regions, separated by a dark gap, known as the Cassini division. On the inner edge of the inner ring is a greyish area. The shadow cast by the rings can be seen faintly on the disc. Apart from the glory of the unique ring system, Saturn is not particularly interesting to look at. It is a heavy yellow in colour, and shines with a dull lustre, which has led to the adjective 'saturnine' as a synonym for cold and gloomy. There are slight signs of belts and occasionally spots on the disc; but they are far less interesting or obvious than the belts of Jupiter. Saturn has ten satellites, and one of them, Titan, is about the size of the planet Mercury. Four others can be seen with moderate telescopes; but the remainder are considerably fainter.

Outwards again from Saturn, we come to Uranus, a planet unknown to the ancients. At its brightest, it can just be seen with the naked eye; but, even with a fairly powerful telescope, it does not show up as much more than a rather dull marble of greenish colour. Though so small to look at, it is, in fact, of great size, and it is only the distance which makes it so insignificant to look at. The chief peculiarity of this planet is that it is 'lying on its side;' that is to say, its axis of rotation is only a couple of degrees from the plane in which it rotates round the sun, compared with 66½° for the earth.

After Uranus comes Neptune, which is a little larger than Uranus, and a good deal more massive. However, being even further away, it is not an object of great interest for those with only moderate telescopes.

Lastly, we come to Pluto, which, as we have noted, is an oddity because it is the wrong size. A theory has been advanced that it is nothing more than a satellite of Neptune, which has

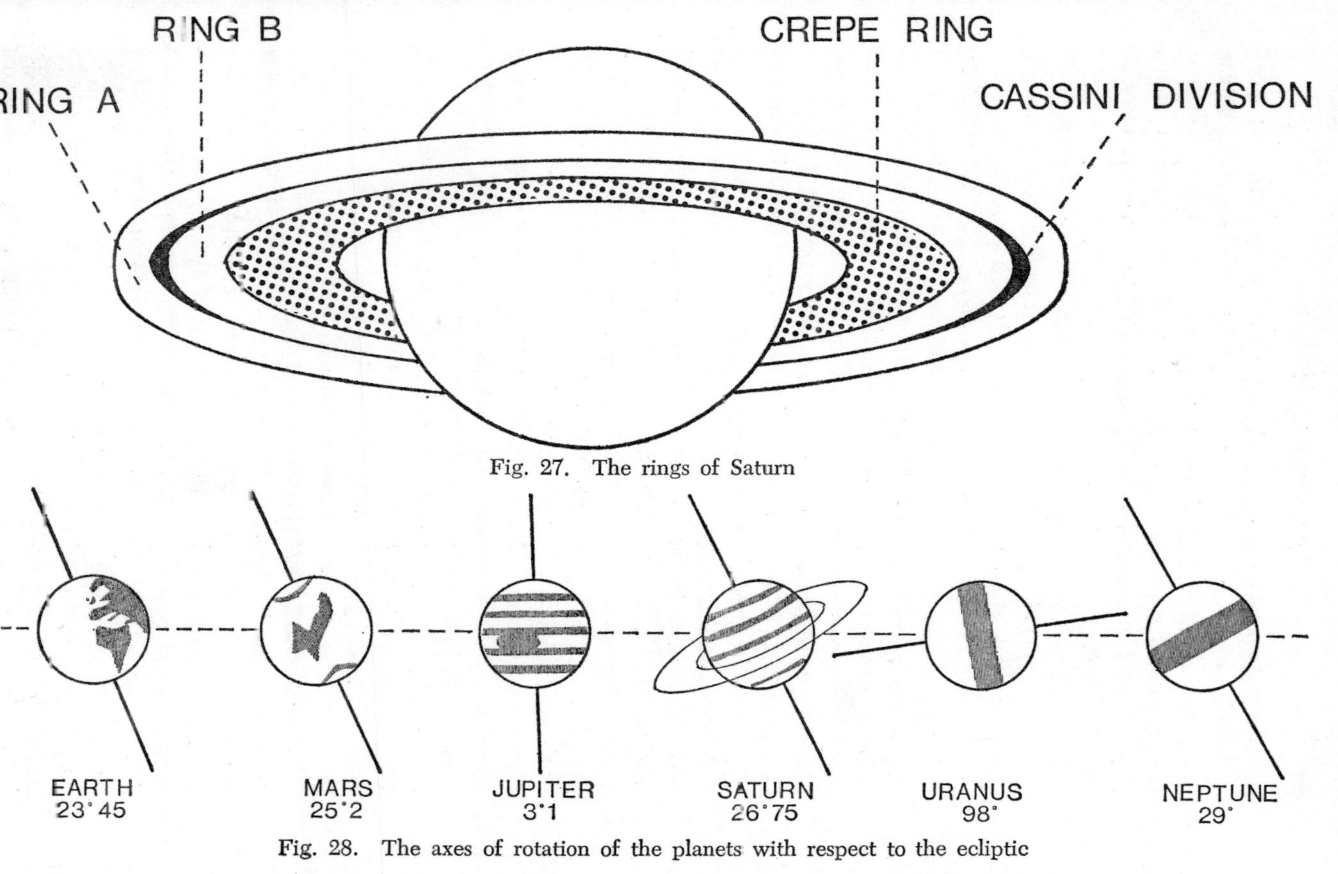

Fig. 27.   The rings of Saturn

Fig. 28.   The axes of rotation of the planets with respect to the ecliptic

broken loose, and taken up an orbit of its own. In any case, it is a very remote, and must be a hostile, world. It is so far from the sun that full daylight there can be no brighter than moonlight on earth. It must also be desperately cold. Perhaps in the next few years we shall learn more about this remote and lonely world.

In addition to the planets, there are two other classes of object which belong to the solar system. The first of these is the comets, which though negligible in mass can provide the most spectacular of all astronomical phenomena. At its brightest, a bright comet can be an awful and wonderful sight, which may even be seen in full daylight, streaming across the sky. It is little wonder that they were regarded as omens by our ancestors. One which appeared shortly before the Battle of Hastings, greatly disturbed the Saxons, who thought it a symbol of disaster. It is not perfectly clear why the Normans did not draw equally gloomy conclusions.

Really large looking comets, visible to the naked eye, are rare, and very many are only known about in astronomical circles. The great majority are only visible through large telescopes, and some do not even have tails. Comets move in highly elliptical orbits; some actually disappear, whilst others come back at regular intervals, though generally long ones. One of the most famous is Halley's Comet, which reappears about every seventy-six years. This is a relatively short period, and other naked-eye comets have periods in excess of a century.

For all their spectacular appearance, even the most famous comets are really trifling objects. The central nucleus is probably nothing more than a collection of ice particles, and has been described as a dirty snowball. Round it is the coma, a haze of thin material. The comet has no light of its own, and can only be seen because of the reflected sunlight. When it gets fairly close to the sun, the solar wind, a stream of particles emanating from the sun, drives off some of the lighter elements of the comet into a streaming tail – by far the most spectacular feature of a naked-eye comet. For this reason, the tail of a comet, when it has one, is always stretching more or less directly away from the sun, and alters direction as the comet circles round the sun at pearihelion.

Not all comets reappear. They may break up when they are

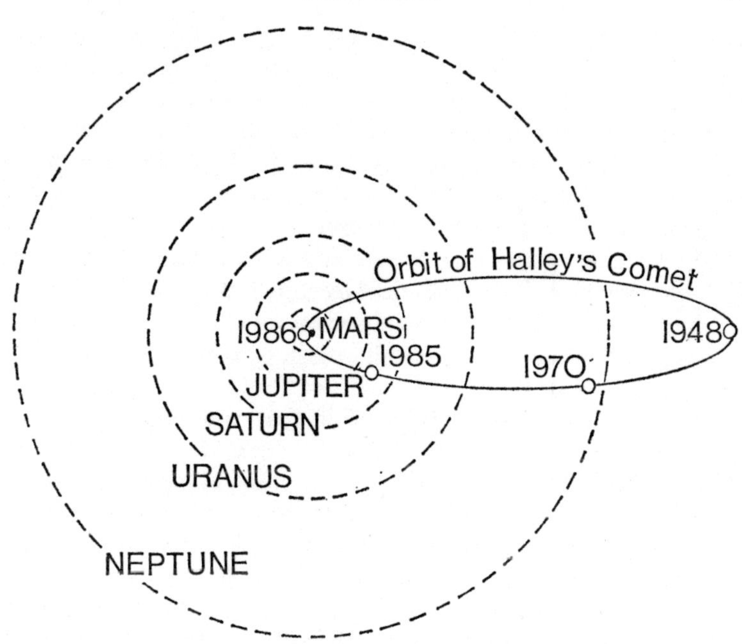

Fig. 29. The path of Halley's Comet

out of our range of sight, which is, after all, the greater part of their lives. One comet, Biela's Comet, broke into two parts, and then disappeared altogether, though at the time when it was due to appear again, there was a bright shower of meteors.

Meteors are the other class of objects which belong to our solar system. On occasions when there is a bright shower, they provide a fascinating sight. Most meteors are wholly insignificant in size, being about as big as a grain of sand. Their brilliance is due to the way they burn up when they enter the atmosphere. Since they are travelling at great speed, the air they encounter on entry causes them to become hot enough to burn, sometimes with great brightness. The heating is due, in part, to the mere friction of the meteor against the molecules of air, but also to the huge compression of the air which cannot get out of the way fast enough. A similar, though more mundane, effect causes a cycle pump to get hot if one pumps hard enough.

Though the vast majority of meteors are very small, and

burn up completely before reaching the ground, there are larger ones which are only partly burned. On very rare occasions a particularly large one strikes the earth, causing a large crater. One such fell in Siberia in 1908. It not only made a substantial crater, but the wind caused by its entry devastated the forest for hundreds of square miles. When a meteor is large enough to hit the earth, it is called a meteorite.

The last meteorite to fall in England was the Barwell meteorite in Leicestershire. This was on December 24th, 1965. It broke in pieces before landing, and many of the fragments were recovered. Probably on entry it weighed around two hundred pounds.

On any clear night it is possible that you may see one or more meteors; but, in general, they occur in showers at fixed dates. This is because a large number are circling in swarms, each swarm in its own orbit round the sun, and showers happen when the earth's orbit passes through the orbit of a meteor swarm. Between July 27th and August 17th one can count on a spectacular display of the swarm known as the Perseids. November 17th may provide another display from the Leonids, though they are less predictable than the Perseids.

The names are derived from the constellations from which the meteors seem to radiate – the point is called the radiant, though this is only an optical effect. The meteors are moving in parallel paths, and create the illusion of coming from a single point, and spreading out as they approach. One can see a similar effect by watching a line of aircraft coming at one from very far away. At a great distance, they will seem bunched up together. Yet, as they come closer, they will seem to fan out. One may pass directly overhead, one away to one's right and a third to one's left.

Comets and meteors are particularly interesting to beginners in astronomy, because the naked eye or a pair of binoculars offer good means of study. Comets and meteors are, however, merely odd scraps of cosmic waste, for all the spectacular displays which they put on from time to time.

CHAPTER VI

The Sun

In the case of the sun, as opposed to the moon and planets, we do have a fair idea of how it came into existence. Since the sun is a very ordinary and average kind of star, we will deal with its creation in the next chapter. One point we might start with is not strictly relevant to the sun as such; but it is a convenient point at which to touch on a question of very wide interest: the question of whether other suns have planets, and whether any of them is inhabited.

As we have seen, the odds are greatly against any other life in the solar system, unless it be some very rudimentary form on Mars. What we do not know is the circumstances in which stars acquire planets. At first sight, it does seem probable that there must be other planetary systems among the myriads of stars. When we consider that we know of something in the order of 100,000,000,000,000,000,000 a hundred million million million stars, it does seem inherently probable that, among so many, there must be a number of stars with planetary systems. Even, however, if there be a great many; even if it be that a planetary system is the rule rather than the exception, it does not absolutely follow that there is sentient life other than on our planet.

We know a certain amount, and guess more, about the conditions which, presumably, led to the creation of life. This knowledge tells us that for life spontaneously to begin there must be a wildly unlikely series of coincidences in the conditions existing in the early days of a planet's history. The odds against proteins being formed by a concatination of circumstances is very large; but, even granted spontaneous protein synthesis, there may have needed to be even more unlikely circumstances which resulted in the formation of the first living cells, even when all the raw materials were present in the primeval environment. Indeed, some biologists would reckon that the odds against the right conditions occurring by chance is of about the same order as the number of all the stars known. On that basis, it would be only about an even chance of life being accidentally created elsewhere.

There is a further consideration to be borne in mind. We would have no chance of directly observing a planet in any other solar system, unless it were of a very large satellite of a very close star. If, then, there be other inhabited planets, we could only get news of it by indirect communication, perhaps by radio messages from other civilizations. But the average distance between suitable stars, even in our own galaxy, is to be measured in many light years. There might be a highly developed civilization in our galaxy whose beings, far more advanced than our own, were now sending out radio messages, which could be picked up and recognized as intelligent signals. Even so, we should not probably receive the signals for a good many tens of thousands of years to come!

Of course the question is open. Even now, efforts are made to search for possible incoming messages and from time to time there is a sensational headline in the newspapers purporting to show that some scientist believes that he has picked up such a signal. Unfortunately, and predictably, it always turns out to be a false alarm.

To stray a little further from the subject, but pursuing an idea naturally associated with astronomy, there is a common misunderstanding which it is worth while to correct. The television networks and the newspapers periodically raise the question of how we might set about sending out signals ourselves, hoping that someone elsewhere would pick them up, or how we would set about communicating with a being from another world who came to visit us.

A little consideration will show that the boot would be almost certainly on the other foot. As we have seen, the world is something under five thousand million years old. Man is anything from, say, thirty thousand to a million years old, according to how one identifies the animal first properly called man. In his history, there has been a period of only a few decades when his knowledge and technology have been sufficiently far advanced for us to contemplate sending out, or recognizing, signals which have been sent in. On this sort of scale, if there be other civilizations with whom it would be possible to communicate, they might be anything from several thousands of millions of years old down to an age comparable with our own. This gives the odds that they would be a more-advanced culture than our own of several million to one. Since only a few hundred years of advance from our own state would imply a technology and scientific knowledge almost infinitely in advance of ours, we may be sure that they would do the communicating. Our part in the transaction would be to listen.

There are other possibilities, though we are straying perilously close to science fiction. It could be that there is a cosmic mechanism which produces life automatically. But, if we study our own history, it could well be that it was a process which, equally automatically, copies what looks like being our own future: that man's natural tendency is for his control over nature inevitably to run in advance of his capacity for understanding his own nature. The preordained cycle of life might thus always end in self destruction. Alternatively, and more hopefully, there could be beings who have reversed the process, and who understand their own natures sufficiently to avoid self-destruction. It could be a slim hope that we might come into contact with such a set of beings, who might yet save us from what we seem determined shall be our fate.

If any excuse is needed for this short excursion to the fringe of our subject, it is that astronomy has always encouraged men to speculate about the vast cosmos in which we live, and space travel has intensified the process.

Now that we are beginning to voyage outside our own home of earth, it would be unnatural if the prospect of visiting even our closer neighbours in space did not unloose a flood of speculation about what we may find in the years ahead.

As we have said, the sun is a very ordinary kind of star. It is only unique to us because it is so close. Ninety-three million miles does not sound all that close, and it takes light around eight minutes to reach us from the sun. The distance is, moreover, between three and four hundred times that of the moon, which, until the last decade, has always seemed to belong to the distant heavens.

It begins to sound closer when we consider that there are some stars so large that, if the sun suddenly grew to that size, we should actually be inside it! Then we need to remember that the nearest star other than the sun is over four light years away. If one drew a chart on a scale which put the sun one inch away from us, the next nearest star would be over four miles distant. To show the further stars in our own galaxy would require a chart thousands of miles across. If we tried to include some of the most remote galaxies...but, at this stage, numbers become useless for the purposes of visualizing. Let us just say that, by astronomical standards, the sun is very close indeed.

It is one of the odd coincidences of astronomy that the sun and moon appear to be almost exactly the same size. But, whereas the moon has only about one quarter the diameter of the earth, the sun is more than a hundred times as wide, and its volume would contain over a million earths. To be exact, the diameter of the sun is 865,000 miles and its mass 1.990×10^{33} grammes. A comparison will show that it is less massive for its size than the earth, having a specific gravity of only 1.4. This is because the sun is composed of gas, and the specific gravity would be a lot less even than this if the gravitational force were not so great that the gas at the centre is very heavily compressed.

It is this huge internal pressure which sets off the complicated process of transmuting hydrogen into helium, giving the sun its vast power of radiation. The resultant temperature at the centre reaches the high figure of 15,000,000°K, which falls off as the surface is approached, until it is only 5,500°. This surface is known as the photosphere; that is to say the visible surface, and consists of high-pressure gas. Above the photosphere there is an 'atmosphere' of much more rarefied gas called the chromosphere. Whereas the photosphere emits light with a continuous spectrum, the chromosphere, being a tenuous gas, absorbs some of the visible light, and causes dark absorption lines in the sun's spectrum. (Fig. 12, page 44)

Though the sun consists for the main part of hydrogen, representing the unused fuel, there are also a number of heavier elements present, including a fair quantity of helium, the main product of 'combustion'. Heavier elements still are also built up in smaller quantities. Indeed, we are probably right in assuming that the original stuff of the universe was hydrogen, and all the heavier elements, such as we find abundantly on earth, are the end products of stellar 'combustion' processes in some, at any rate, of the stars.

The sun's disc is not without features, though they are transitory. The most notable of these are the sunspots, which appear as black patches, though, in fact, that are only apparently dark because they are less brilliant than the rest of the photosphere – perhaps 1,500 or 2,000° lower in temperature. A spot is generally graded in darkness, with the centre, or umbra, darker than the surrounding area, or penumbra. The actual shapes of the spots are complex and variable; moreover, they tend to come in groups, which can be seen moving across the disc as the sun slowly rotates, about once in three and a half weeks. Some of the spots endure long enough to be seen and recognized after disappearing behind the limb of the sun, and reappearing about a fortnight later.

The actual nature of a sunspot is a vortex of the gas which composes the surface of the sun. It has been discovered that they are associated with strong magnetic fields, and magnetic forces are probably concerned in the birth of sunspots.

Though no one quite knows why, sunpots come in cycles of maxima and minima, going through a complete cycle in about eleven years. There are still some spots at the time of minimum activity, but, on average, far fewer than at the time of maximum activity. The recent maxima have been in 1947, 1957–8 and 1968, though the last of these produced fewer spots than usual. There will be another maximum in about 1979.

Sunspots do not represent the whole of the activity on the surface of the sun; indeed, the whole outer body is in a state of constant turmoil, though this is largely invisible from the earth. One activity closely related to sunspots are the so-called faculae, from the Latin word meaning torches, which appear as bright patches. These faculae often appear in a place where a sunspot will develop, and persist after the sunspot has vanished.

From time to time we are presented with an opportunity to see features of the sun which are normally invisible. This is

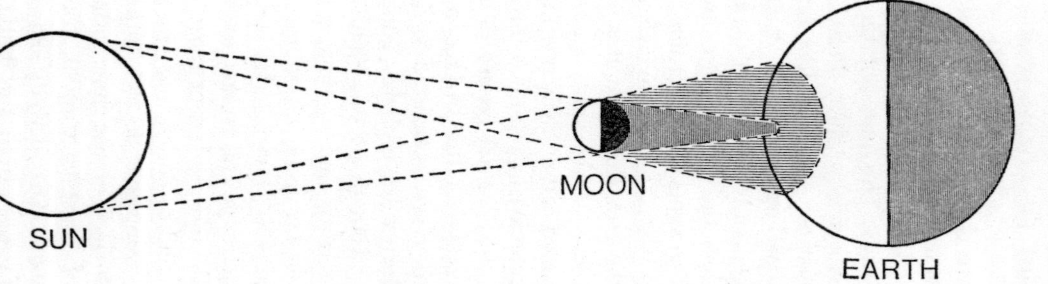

Fig. 30.   The respective sizes of the earth and the sun

when the moon passes directly across the face of the sun in a total eclipse. The extent of times during which an eclipse is total varies according to the relative distances of the sun and moon at the time when the eclipse takes place. Totality is only observable over a short and narrow band of the earth, because the full shadow of the moon only just touches the earth, as the drawing will show. The period of totality may be only a few tens of seconds, and cannot be longer than seven minutes.

Since the disc of the moon just about covers the surface of the sun, there is a chance for the relatively darker outer layers of the sun to become visible, especially the very diffuse corona, or outer 'atmosphere', which stretches for millions of miles beyond the photosphere. A total eclipse can be a breathtaking sight, and it is not to be wondered at that people travel great distances to witness one.

Among the things to be seen during a total eclipse are the prominences, once known as red flames – which was a bad name for them, because they are not flames, but masses of red glowing hydrogen. These can be divided into two classes: eruptive and quiescent. As the names suggest, the quiescent prominences are fairly static, and can remain visible for quite long periods. The eruptive prominences, on the other hand, are dynamic occurrences, and can be rendered dramatic by films taken with exposures at short intervals. More recently, new instruments have been devized which enable astronomers to view or photograph otherwise invisible events on the sun by using combinations of narrow moving slots and a prism or grating which produces a spectrum. The result is that the disc can be viewed in a single wavelength of light at any chosen length. These instruments are known as spectroheliographs or spectrohelioscopes, according to whether they are used for photography or visual viewing. Since the light emitted by the different outer layers of the sun differs from one another, any part can be examined by choosing to view at the appropriate wavelength.

We have said that the corona of the sun stretches for millions of miles; but it is difficult to choose any distance at which it can be said to end. Indeed, in some respects, we might almost say that the earth comes within the corona. The physics of the sun, in fact, is becoming very complex indeed. It is constantly emitting streams of both radiation and particles, some of which become trapped in the magnetic field of the earth. The now famous Van Allen belts, first discovered when satellites and probes were sent out into the area, consist of such trapped particles. Others are deflected by the magnetic field to the polar regions of the earth, and are associated with the displays of aurora, visible only in the high latitudes. A disturbance on the sun is often followed a day later by displays of the aurora and magnetic storms, sometimes associated with radio blackouts. The delay of a day represents the time taken for the charged particles to reach the earth at about 1,000 miles a second.

The sun, which is the author of our own life and being, is a fruitful source of interest to the amateur astronomer, though the opportunities for observation are chiefly associated with tracking and drawing sunspots. It is, however, becoming a highly recondite subject, and anyone wishing to take up this branch of the subject will need to study in much greater detail the physical processes involved. One warning must be repeated here. *On no account should anyone try to look at the sun through a telescope or even weak binoculars. To do so is to invite instant blindness.* It is even highly unsafe to use smoked glass, or optical filters at the eyepiece. These may quite easily be cracked without warning by the heat, with catastrophic results. In the chapter on equipment and observing the matter will be dealt with at greater length.

CHAPTER VII

The Galaxy

Look up into the sky on a dark, clear night, and you will see
a great many stars – not so many as might be thought at first
glance, but several thousands. You will also see something else:
a luminous band, stretching from one side of the sky to the
other. This is the Milky Way. It is important to appreciate just
what this is, because it is the key to the whole idea of our star-
system or Galaxy.

A star, as we have seen, is a sun. In our particular system
there are about 100,000 million suns. They are not spread about
at random; they form a somewhat flattened system, as shown

Fig. 32.   The position of the sun in the Galaxy

in the diagram, with the sun well away from the centre. (On
this scale, it is permissible to say that the sun and the earth
are in the same position.) Look along the main plane of the
system, and you will see many stars in much the same direction;
and this is what makes up the Milky Way band. Look at right
angles to the main plane, and there will be far fewer stars.

The significance of this is that the stars in the Milky Way
are not crowded close together, even though they may appear
as though almost touching each other. Each star is a very long

way from its neighbours, unless we include the physically connected pairs or binaries which will be described later (page 110).
Our star system is a big place. It would take a ray of light
100,000 years to pass from one end of it to the other, and 20,000
years to cross the thickest part of the central bulge or galactic
nucleus.

Actually, no ray of light could slice straight through the
whole system. The Galaxy contains a great deal of thinly-spread
material, and this material absorbs light, so that we cannot
see even as far as the centre of the system; our view is blocked.
Luckily, the long-wavelength radio waves are not so hindered,
and now that radio astronomy has become a major science we
can at last learn something about those regions which we can
never see visually.

The scale of the universe is something that is rather too hard
to understand. It is possible to draw some idea of the distance
of the moon (less than a quarter of a million miles – a mere
ten times round the earth's equator) and even the sun and
planets; but the stars are almost inconceivably remote. It was
hardly surprising that the ancient astronomers had no idea of
how large the Galaxy really is, and it was not absurd for them
to suppose that the stars were luminous points fixed to a solid
crystal sphere. In fact, it was only in 1838 that a German
astronomer, F. Bessel, managed to measure the first star-distance.

His method was straightforward in theory, though not in
practice. He selected a star which he thought might be a close
neighbour; this was 61 Cygni, in the constellation of the Swan.
Then he measured its position with reference to adjacent stars
at a six-monthly interval, as shown in Fig. 23, (which has
been drawn wildly out of scale, to make it clear). Obviously,
there was an apparent shift in position, due to parallax. Bessel
measured the shift and deduced that 61 Cygni must be about
eleven light-years away from us. He was correct; and since
this was in the pre-photographic era, and he had to carry
out all his work visually, his was a remarkable triumph.

This method, trigonometrical parallax, can serve for stars up
to a few hundred light-years from us, but stars which are
further away show parallax shifts which are too small to measure.
Less direct methods have then to be used. Most of them, though
not all, involve finding out how luminous the star really is, and

a, showing the Straight Wall. Photographed by Commander Henry
Hatfield, with 12″ telescope and camera both made by himself

Moon, Schickard area, photographed by the author with convex lens

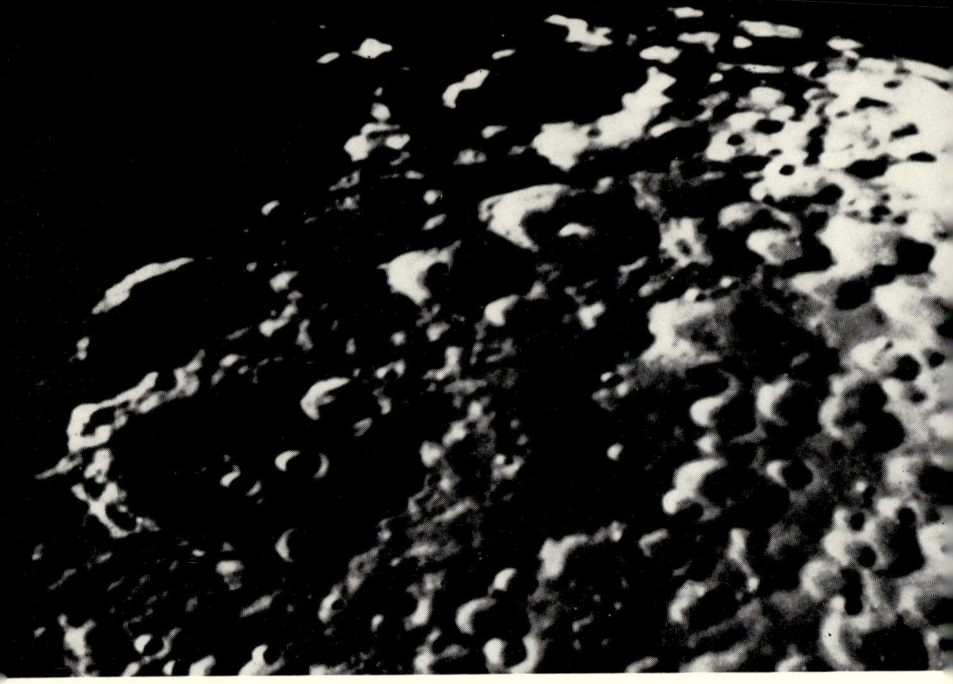

Above Moon, area of Clavius, photographed by the author with a convex lens. *Below* Quarter moon, photographed by the author with single concave lens. The Straight Wall can just be seen, also the crater Clavius. Compare with the larger one of this crater taken with a convex lens

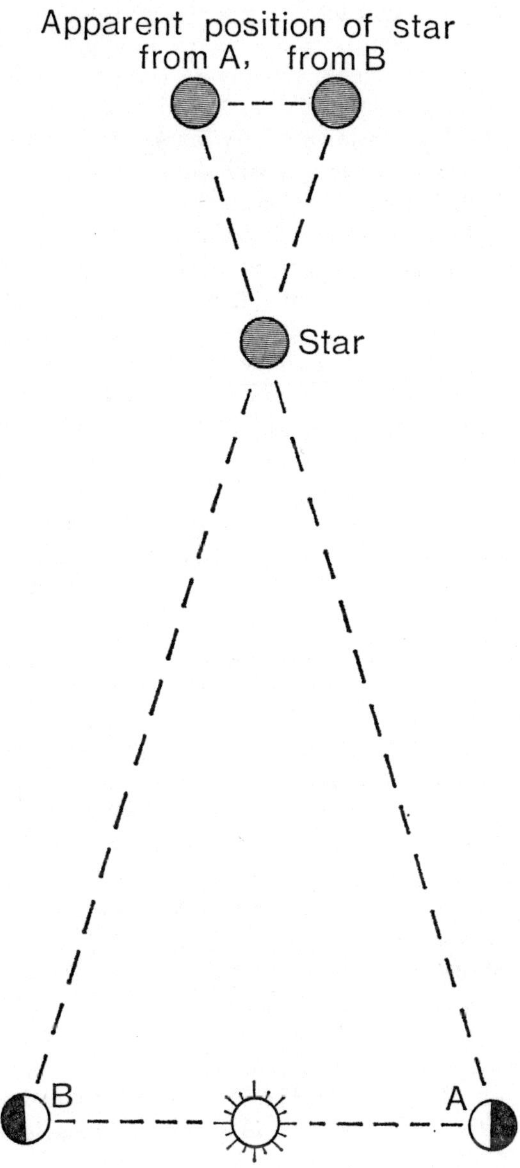

Fig. 33. Shift in parallax of distant star seen from the opposite ends of the earth's path round the sun A and B. The further star will seem to move behind the closer star

D

then comparing the real brilliance with the apparent brightness; the distance can be computed – provided, of course, that essential refinements are introduced, notably to compensate for the absorption of a star's light by interstellar dust and gas.

In the early days, little could be found out about the nature of a star, because no telescope will show a star as anything but a point of light. Neither are there any obvious movements. The stars are not fixed in space, and the old term 'fixed star' is a gross misnomer; but they are so remote that their individual or proper motions are very slight, and cannot be noticed with the naked eye even over periods of many lifetimes. Admittedly the celestial pole is in a different place, but this has nothing to do with the stars themselves. Faced with an array of motionless twinkling points, the astronomer of former times was obviously limited.

He could, of course, see that the stars are not all the same. In the magnificent constellation of Orion, for instance, there are two particularly brilliant stars, Betelgeux and Rigel. Betelgeux is orange; Rigel is white – and it takes no genius to deduce that Rigel is the hotter of the two; white heat is hotter than orange heat. But which is the larger star? Nowadays we know that Betelgeux is much the bigger of the two; but before its distance could be measured, there could be no way of knowing.

Learning one's way around the sky is not nearly so difficult as it might seem. The best plan is to select a few obvious constellations, and use them as signposts to locate the rest. Consider, for instance, the celebrated Great Bear (Ursa Major) which is often called the Plough, though to be strictly accurate the nickname applies only to the seven most famous stars in the constellation. Two of the Bear stars, known as Dubhe and Merak, point almost directly to Polaris, the Pole Star, in the much fainter constellation of the Little Bear (Ursa Minor); the curve of the Great Bear's tail leads to a brilliant orange star, Arcturus in Boötes (the Herdsman); Orion's belt points one way to the orange Aldebaran in The Bull (Taurus) and in the other direction to the brightest star in the sky, Sirius in the Great Dog (Canis Major) – and so on. A simple set of star maps, together with a few evenings' practice, will work wonders.

There are two important points to be noted here. First, there is the term 'magnitude', which is a measure of apparent bril-

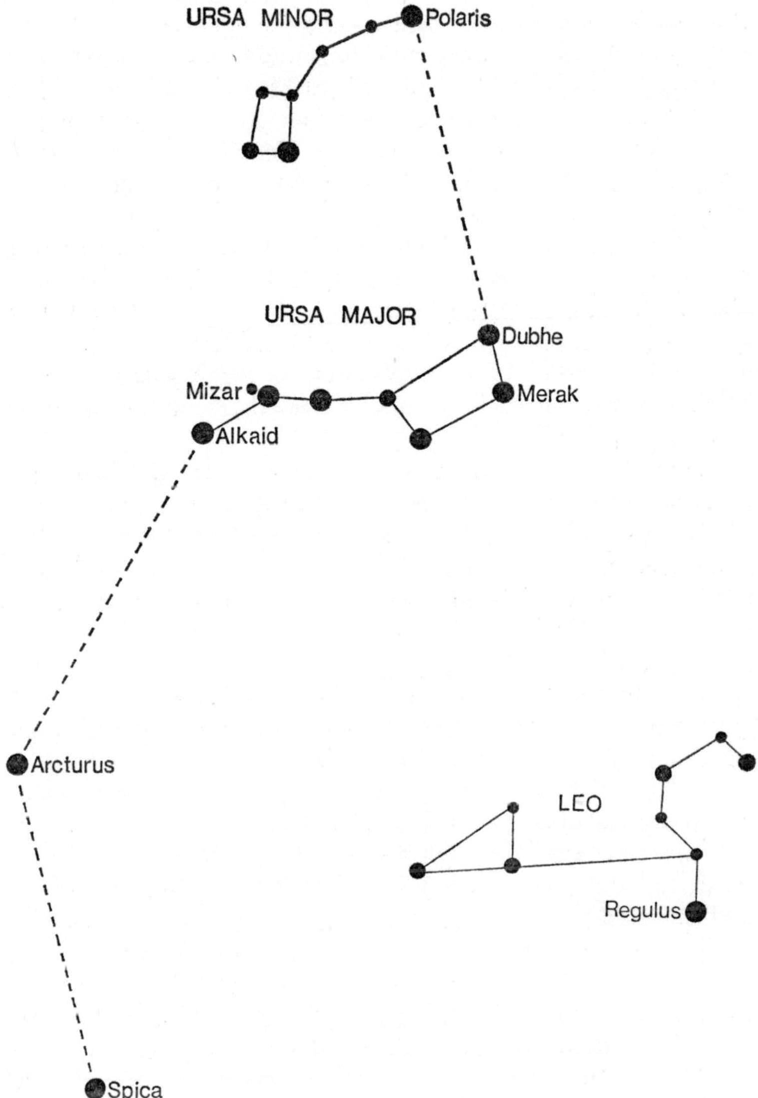

Fig. 34.   Finding the Pole Star from the Great Bear (Ursa Major).
A few other stars are shown

liancy. This works in the manner of a golfer's handicap. The brightest stars have the lowest magnitudes; thus Aldebaran (magnitude 1) is brighter than Polaris (magnitude 2). The stars range in brightness from Sirius, magnitude *minus* 1.4, down to dim stars below magnitude *plus* 20, visible only with the world's greatest telescopes. There are only four stars – Sirius, Canopus in the Ship, Alpha Centauri in the Centaur, and Arcturus – above magnitude 0, though several others are only just below. Rigel, for instance, is +0.2.

A star's apparent magnitude is not a reliable key to its real luminosity. It may shine brilliantly only because it is near. Sirius is a case in point. True, it is 26 times as powerful as the sun; but Rigel, which is not nearly so bright in our skies, is 50,000 sun-power! The explanation is that Sirius is a mere 8.6 light-years away; Rigel lies at something of the order of 900 light-years.

Secondly, a constellation is not an entity. In the Bear's tail there are two stars, Mizar and Alkaid, which seem to lie side by side. (See Fig 34) Yet they are not neighbours; Alkaid is the more luminous and more remote of the two – and in fact it lies at well over twice the distance of Mizar, so that Alkaid is further away from Mizar than we are. It is all a question of line-of-sight effects. The constellations were arbitrarily drawn up in ancient times, though others have been added since until our modern maps show 88 of them. There is nothing at all significant about the patterns – or the names. Also, it takes a great effort of imagination to twist the patterns into the shapes which are associated with them. The Great Bear is more like a saucepan; Pegasus, the Flying Horse, is a square (in the literal sense, not the vernacular); Virgo, the Virgin, resembles a large and distorted Y.

The great breakthrough in stellar astronomy came during the 19th century. We have already said something about the spectrum of the sun, which is made up of a continuous rainbow crossed by dark absorption lines. At an early stage in astronomical spectroscopy it was found that normal stars have spectra of the same kind, but they are not identical; the spectrum of a red star such as Betelgeux differs markedly from that of a yellow star, such as Capella, or a white star, such as Rigel. Since each dark line represents one particular element, or group of elements, it became possible to find out what substances were present in the stars. The essential clues were found by G. Kirchhoff of Heidelberg in 1859; by the end of the century the spectra of many stars had been studied and catalogued.

Something strange emerged. Orange and red stars came in two varieties; very luminous (giants) and very faint (dwarfs). There was a notable but less marked difference for the yellow stars, but not for the hot white and blue ones. Two astronomers, H. N. Russell in America and E. Hertzsprung in Denmark, investigated the whole problem, and produced the first example of what is now called a Hertzsprung-Russell or H-R diagram. This particular piece of work proved to be of absolutely vital importance.

In an H-R diagram, a star's spectrum is plotted against its luminosity. (One way of expressing luminosity is by absolute magnitude – that is to say, the apparent magnitude that a star would have if it could be seen from a standard distance of 10 parsec or 32.6 light years: the sun would show up as a dim naked-eye star of magnitude 4.9, whereas Rigel, at magnitude −7, would cast shadows.) It is obvious at a glance that the greatest number of stars lie on a band stretching from the upper left to the lower right; this is called the Main Sequence. The letters in the 'spectrum' column indicate spectral classes; O, B and A for white stars, F and G for yellow, K, M, R, N and S for orange and red. This rather chaotic sequence was a result of several original errors in classification, but it has become so firmly established that it will certainly be retained; of these classes stars of types O, R, N and S are relatively uncommon, and the principal series is from B to M. It is, of course, a series of decreasing surface temperature.

The sun is a typical Main Sequence star; it is yellow, and comes under type G. Rigel is type B; Sirius, A. Arcturus, K; Betelgeux, M; and so on. However, Arcturus and Betelgeux do not lie on the Main Sequence. They are included in a less populous but still well-marked Giant Branch. The sun is a yellow dwarf; Capella in the Charioteer (Auriga), far more luminous, is a yellow giant. There are no red stars which are equal in luminosity to the sun. All the red stars are either much more powerful, or much more feeble: searchlights or glow-worms.

Astronomers in general were excited by H-R diagrams, and

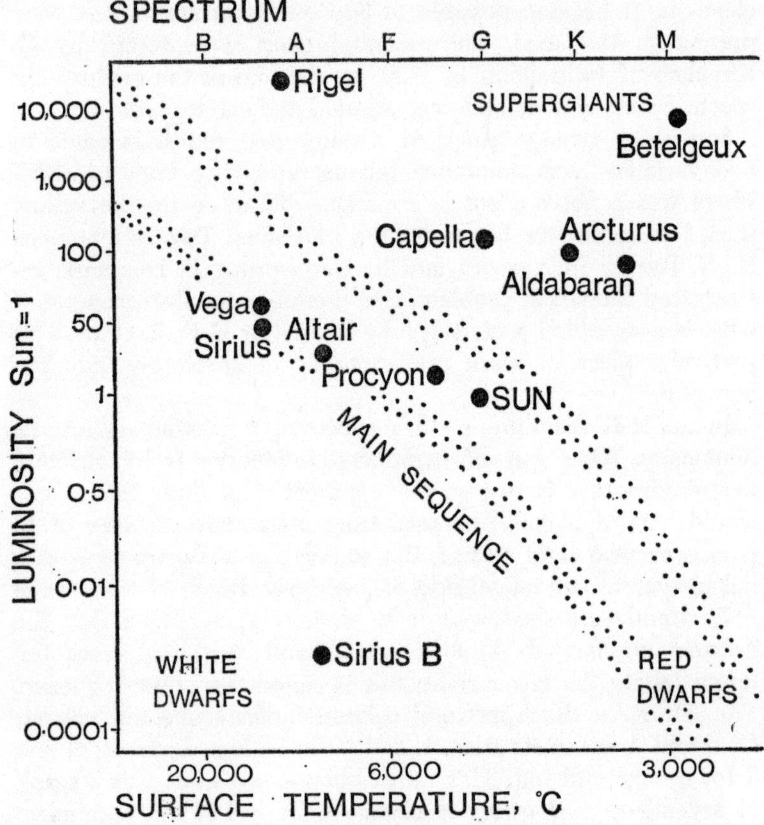

Fig. 35. Typical H–R Diagram

tended to regard them as pointing out an evolutionary sequence for an average star. This sort of view prevailed until
well into the inter-war period. It all seemed so delightfully
simple. A star began by condensing out of a gas-and-dust cloud
or nebula, of which there are plenty in the Galaxy. It shrank,
by gravitation, and hotted up. Energy production began inside
it, by nuclear reactions of some sort, and the star became a
young red giant (Betelgeux), turned yellow and then white,
joining the Main Sequence to the upper left. It was then at its
peak. As it continued to shrink, its luminosity fell away; it passed
down the Main Sequence, and ended up as a red dwarf before

all its light and heat departed, leaving it as a stellar corpse.

Alas, in astronomy – as in so many other things – matters are seldom straightforward. As time went by, disquieting facts began to emerge. First, what made a star shine? Certainly not ordinary combustion. A star made up of coal, shining as brightly as the sun does now, would burn away in a few million years at most; but we know that the sun is at least 5,000 million years old. The direct annihilation of matter, in which atoms were mutually destroyed and energy released, gave a time-scale, which was just as obviously too long. Finally, just before Hitler's war, the answer was found – and it altered the whole concept of the significance of the H-R diagram.

Basically, what happens is, as explained in Chapter II, that the hydrogen inside normal stars is being changed into helium. The sun functions in this way, and, as we have noted, it is losing 4,000,000 tons a second. Very energetic stars, such as Rigel or even Sirius, are much more spendthrift. They will have used up their available hydrogen 'fuel' long before the time, perhaps 8,000 million years hence, when our mild sun begins to run short.

Now let us go back to the H-R diagram, and start again. We can begin, as before, with a star condensing out of a nebula, but after that we must jettison all our previous ideas. The star contracts, and as soon as its interior has reached a few million degrees in temperature the hydrogen-into-helium process can begin. The star has joined the Main Sequence – but where it joins the line depends primarily upon its original mass. If it is very massive, it will join near the top left (as Rigel has done). If it is less massive, it will join near the middle (as the sun has done). If it is of low mass by stellar standards, it must be content with the lower right.

This is where the star spends most of its active career as a luminous body. If it is feeble, it will radiate gently and will survive on the Main Sequence for a long time; if it is energetic and massive, it will have a briefer spell there – in the case of a star such as S Doradus, which is at least a million times as powerful as the sun, perhaps only a few thousand years. Gradually, a core of helium will be produced, and the reactions will go on in the hydrogen shell outside the core. But at last the crisis will come. The hydrogen supply will fail.

A star has always to balance itself; the gravitational force tends to pull it inward, the radiation pressure from inside tends to distend it. In a Main Sequence star these opposing forces cancel each other out. But when radiation virtually stops, gravitation must take over; and the star begins to shrink. This heats up the helium core, and new nuclear reactions begin – with what is called the 'helium flash'. No longer is the helium inert; it too starts to build up more complex elements. The temperature at the core rises far beyond the 15,000,000 degrees or so of a Main Sequence star. Heavier elements are now being synthesized.

We cannot pretend to be confident about the sequence of events after the helium flash, but at any rate it seems that various kinds of reactions take place until there is a new crisis; the nuclear power is used up. Meantime, the star has become a red giant, with a fiercely-hot energy-producing core, where the heavy elements are being made, and a swollen shell around. It may be that at last the outer shell is thrown off altogether, leaving only the core; the result is what is called a planetary nebula (though let it be added that not all astronomers are agreed about this). Very massive stars may explode, because their nuclear reactions have 'run away;' these vast outbursts do occur sometimes, and are called supernovae. But in most cases the core of the star is left as a very small, very dense body. The atoms in it are broken up, and the component pieces are jumbled together with almost no waste space. The star is shorn of its glory; it has become a White Dwarf.

White Dwarfs appear at the lower left of the H-R diagram. They are in no way comparable to the red dwarfs. A White Dwarf is a bankrupt star, devoid of nuclear energy, and shining feeble even though its surface is still hot. Because of the packing of the shattered atoms, the density is remarkable – one could cram many tons into a thimble. The best-known example is the faint companion of the brilliant Sirius, where the diameter is a mere 24,000 miles (smaller than the planet Uranus) but the mass is equal to that of the sun. A few known White Dwarfs are equally massive, but even smaller than Mars. They are indeed peculiar objects. Undoubtedly they are very common; but because they are so faint, we can detect only the closer ones.

This picture is very different from the original one! Instead of being youthful, red giants prove to be ancient; they are approaching death. Yet their death must be protracted. A White Dwarf can continue radiating feebly for thousands of millions of years. What finally happens to it is uncertain. Recently radio astronomers have detected some remarkable sources which are 'pulsing' in the radio range, and it is thought that these pulsars as they are called, are neutron stars – made up of neutrons formed by the jamming together of protons and electrons. It may be that pulsars mark the end of a star's career as an active body. One pulsar has been optically identified with a dim, flashing object; and since this lies in the gas-cloud known as the Crab Nebula, it is very significant.

The Crab Nebula, visible with a small telescope near the star Zeta Tauri in the Bull, is the wreck of a supernova seen by Chinese astronomers in the year 1054. For some months the supernova was bright enough to be seen in broad daylight, and must have been radiating at least ten million times as fiercely as the sun. Then it faded from sight; but there can be no doubt that the Crab Nebula is its débris. The gas is still spreading outward from the old explosion-centre. What the Chinese stargazers saw was nothing more nor less than the destruction of a sun; but as the Crab is 6,000 light years away according to the latest estimates, the explosion really took place 6,000 years before A.D 1054.

The Crab sends out radio waves as well as visible light. It also emits X-rays, and now we have the pulsar to add to the scene. Because it is the most prominent supernova wreck, it is of immense importance to astronomers. One eminent scientist has commented that there are two branches of astronomy; the astronomy of the Crab Nebula, and the astronomy of everything else!

Let it be repeated that this picture of a star's evolution is not conclusive. We are fairly happy about the middle chapters of the story, and reasonably so about the beginnings, but we are not at all confident about the closing sections. Yet more knowledge is being gained all the time – and, undoubtedly, we know a great deal more than we did only a decade ago.

There are other kinds of stars to be considered. Quite apart from supernovae (of which only four have been seen in our Galaxy in historical times; the last was in 1604) there are ordinary novae, in which a faint star suddenly flares up to many

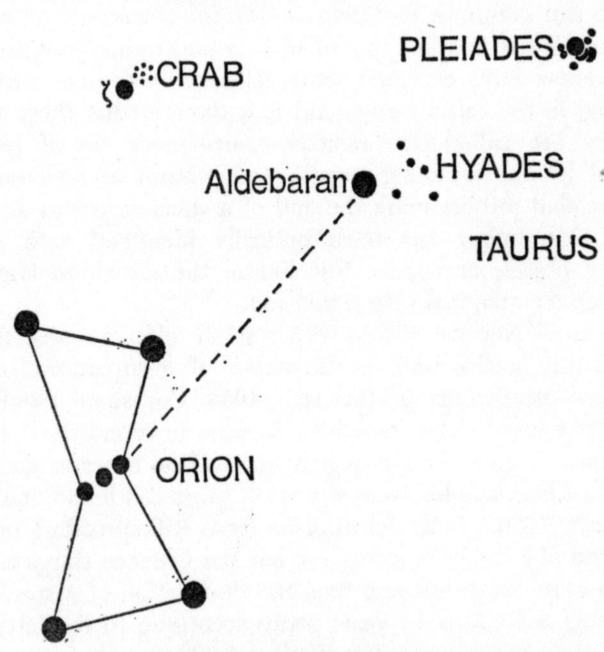

Fig. 36.　Orion and Taurus

times its usual brightness before fading back to its former obscurity. Here, the outburst seems to affect only the star's outer layers, and there is no cataclysmic explosion of Crab Nebula's violence. Many novae have been seen. The most interesting of modern times have been those of 1901 and 1918 (each of which rivalled Sirius for a while), 1934 (in Hercules – which equalled Polaris) and 1967 (in the constellation of the Dolphin, Delphinus). Of these, the last was discovered by the English amateur astronomer, G. E. D. Alcock, who makes a hobby of hunting for both comets and novae, and has achieved remarkable success. The novae remained visible to the naked eye for some time, and at the time of writing (January 1970) it was still to be seen in binoculars; before its outburst it was of the 12th magnitude, and no doubt it will revert to this modesty in the long run.

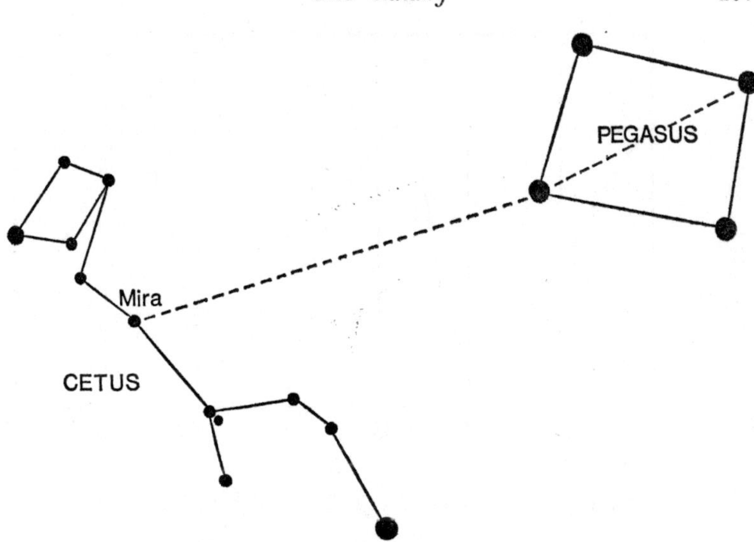

Fig. 37.   How to identify Mira

Associated with novae are the variable stars, which are exactly what their name implies. Unlike stable stars such as the sun, the variables brighten and fade over reasonably short periods. They are pulsating, and their diameters as well as their luminosities change. Amateur astronomers do invaluable work in keeping check on them; today, variable star research is undoubtedly one of the most important branches of amateur astronomy.

Some of the variables – mainly red giants – have relatively long periods; thus Mira, in the Whale (Cetus) comes to maximum once every 331 days or so. It is then visible with the naked eye, and sometimes, as in 1969, it equals Polaris, though at minimum it is difficult to see with binoculars, and a telescope has to be used. Others are irregular in their behaviour; we never know quite what they will do next – for instance R Coronae, in the Northern Crown, is usually on the fringe of naked-eye visibility, but sometimes sinks to below the 13th magnitude, staying at minimum for some time before climbing back to normality. But it is the Cepheids which are of the greatest interest to modern researchers.

Cepheids take their name from Delta Cephei, the brightest

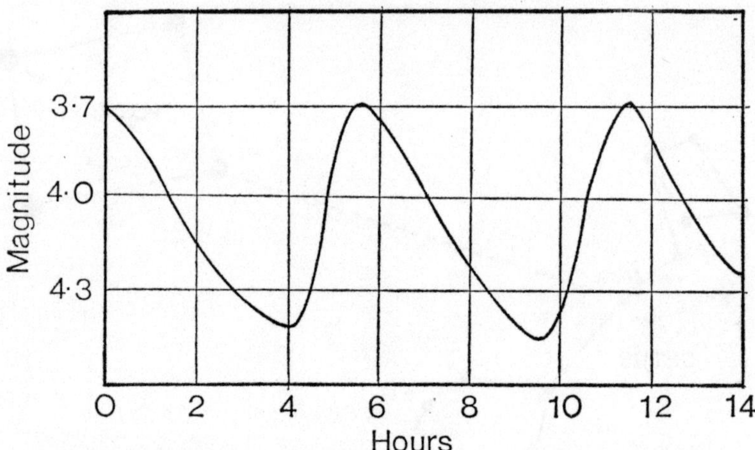

Fig. 38.   Luminosity curve of a Cepheid variable (Delta Cephei)

member of the class. Their magnitude ranges are not very great (Delta Cephei itself changes by only a magnitude) and their periods are regular; Delta Cephei takes 5.3 days to pass from one maximum to the next, while another naked-eye variable of the same type, Eta Aquilae in the Eagle, takes 7.1 days. The important point is that the luminosity of a Cepheid is linked with its period of variation; the longer the period, the greater the luminosity. Thus Eta Aquilae is more powerful than Delta Cephei. Since it looks almost equally bright, it must clearly be more remote – and in fact the Cepheids 'give away' their distance for nothing. Once the period has been measured, the luminosity can be found; this gives the distance. The principle is much the same as that of estimating the distance of a ship's light out to sea. If you know that the light is very powerful, and yet it still looks dim, then the ship must be a long way off.

Cepheids are powerful stars, and can be seen over great distances. They are found in other galaxies as well as our own, which is very lucky for us – otherwise, we might still be groping around trying to decide whether or not our Galaxy could be the only one. Associated with Cepheids, but not identical with them, are the so-called RR Lyrae variables, which have still-shorter periods and seem to be of more or less uniform luminosity, so that they too can be used as 'standard candles'.

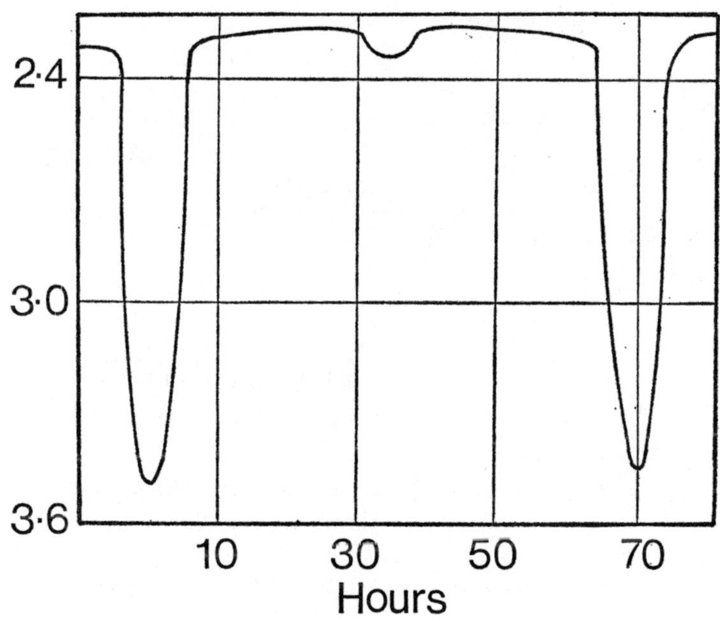

Fig. 39.   Luminosity curve of an eclipsing binary (Agol)

There are, then, variables of all kinds; but there are some stars which seem to brighten and fade, yet are not truly variable at all. The classic example is Algol in Perseus, which is normally of the second magnitude. Every two and a half days it fades down to magnitude 3½, and takes some time to recover; it seems, in fact, to 'wink' slowly and regularly. As long ago as 1783 John Goodricke, a deaf-mute astronomer who lived for a tragic-ally short period of 21 years suggested that this behaviour of Algol was due to the fact that the star is made up of two components, one bright and one dimmer, moving round their common centre of gravity in a period of two and a half days; *when* the faint star passes in front of the brighter, and hides or eclipses it. Goodricke was, of course, quite right, and many eclipsing variables are now known. Not all are the same as Algol; in many cases the two components are not very unequal, so that there are two minima of different depth. Beta Lyrae, near the brilliant blue summer star Vega, is one such. The two components of Beta Lyrae must be so close together that they

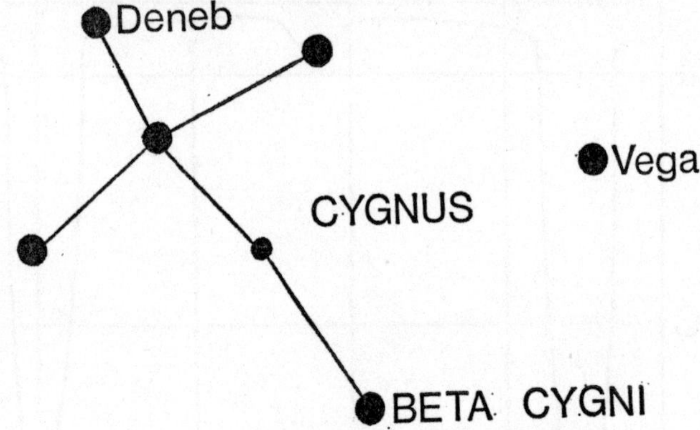

Fig. 40.   The position of the star Beta Cygni

almost touch, and each is drawn out into the shape of an egg, while there is a common gas-cloud around them. Needless to say, this picture has been drawn up theoretically; to us, Beta Lyrae, looks like any other star – apart from its changes in magnitude.

There are many other pairs of stars in the sky in which there are no eclipses as seen from earth. In many cases the revolution periods are extremely long. These double stars are called binaries; the most famous is Mizar, the second star in the tail of the Great Bear, which is easily identifiable because it has a much fainter star (Alcor) close beside it. Telescopically, it is found that Mizar is itself made up of two components, so close together that to the naked eye they appear as one. Other famous binaries are Gamma Leonis in the Lion, Gamma Virginis in the Virgin, and the brilliant southern star Alpha Centauri. A fainter companion to the Alpha Centauri system, known as Proxima, is a mere 4.2 light years from us; apart obviously from the sun, Proxima is the closest of all the stars – but it is a dim red dwarf, much too faint to be seen with the naked eye.

Binaries are spectacular objects as seen with a small telescope, and some of them are gloriously coloured. Undoubtedly the best example is Beta Cygni, the faintest of the five stars making up the cross which is known as Cygnus, the Swan. (It is at its best during summer evenings when, as seen from Britain, it is practically overhead.) Beta Cygni is made up of a yellow primary

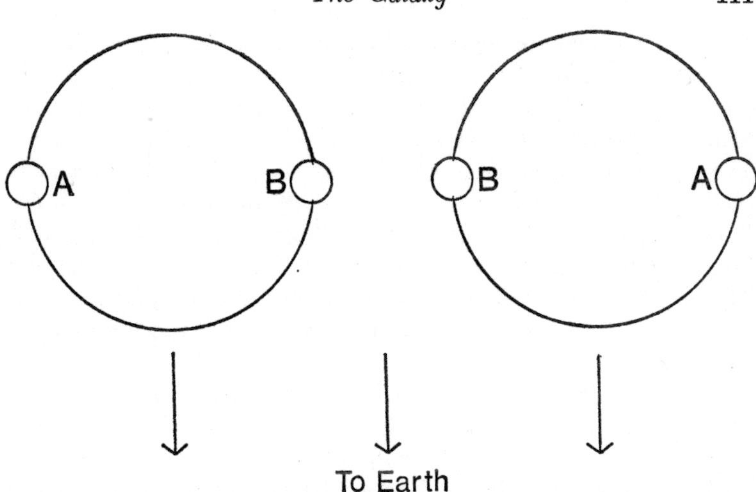

Fig. 41.   On the left hand A is approaching and B receding. On the right, the positions are reversed

together with a companion which some people describe as blue and others as emerald green; either way, it is well worth looking at. A very small telescope will show it well.

With Beta Cygni the separation between the two stars is very wide (over 30″ of arc as seen from earth), and so their period of revolution round their common centre of gravity – the barycentre in the earth-moon system – is extremely long. Other binaries are close together; for instance the fine pair Zeta Herculis, where the primary is of the 3rd magnitude and the companion between the 6th and 7th, has a period of only 34 years, so that the changes in the position angle – that is to say, the direction of the fainter star relative to the brighter – can be noticed over reasonably short periods.

There are many double stars in the sky. Oddly enough, most of them are binaries. One can have the situation in which two stars seem close together even though one is much further away than the other merely because they lie in the same direction; but such cases are surprisingly rare. One of them is Vega, the brilliant blue star which is so prominent during summer evenings. Vega has a 12th-magnitude companion which is not associated with it, and simply happens to lie in almost the same direction as seen from the earth.

If the two components of a binary are so close together that no telescope can show them separately, we can still find out the dual nature of the star. If both components are bright, both will produce spectra. The revolution period is bound to be short. There will be times when one component (A) is moving towards us and the other (B) away, as in the first diagram (Fig. 41); A's lines will therefore be blue-shifted (Doppler effect) while B's will be red-shifted. Half a revolution later, the opposite will apply. The spectral lines will perform a rhythmical dance, from which astronomers can extract all (or, at least most) of the information they need.

There are star-families in which three, four or more components may be seen. Castor, in the Twins, is made up of two bright stars separable in a modest telescope; each component is a spectroscopic binary. Some distance away is the much fainter Caster C, also a spectroscopic binary. Therefore, what we see with the naked eye as one star is really six. Multiple stars are not uncommon. Anyone living on a planet moving round a component of a multiple star would be presented with a most interesting sky, but also, one must admit, with a somewhat erratic climate.

Going on in logical sequence, we progress from the multiple stars, with a few components, to the star clusters, with hundreds. The best example of a loose or open cluster is the Seven Sisters, more properly known as the Pleiades, which lies in Taurus (the Bull) and is known to anyone who has taken only a passing interest in astronomy. Seven stars are visible to the naked eye on a clear night, and binoculars show dozens; the leading members of the cluster are hot and white, and it is almost certain that they had a common origin. Mixed in with the stars is a great deal of thin gas, revealed by long-exposure photographs taken with large telescopes.

The other celebrated open cluster is the Hyades, round Aldebaran and yet Aldebaran is not a member; it lies between the Hyades and ourselves, which is a striking demonstration of the line of sight effect which makes us think of the sky in two-dimensional terms. There are, of course, a large number of open clusters, some of which can be seen without a telescope; and there are the much more regular globular clusters, which lie round the edge of the Galaxy. From Britain, only one of these, the cluster in Hercules, is a naked-eye object, but there are two brighter ones in the far south of the sky.

It was the globular cluster research which provided us with the key to the size of the Galaxy. They contain RR Lyrae variables which, as we have noted, give away their distances merely by varying light. The American astronomer Harlow Shapley studied the RR Lyrae variables, measured the distances of the globular clusters in which they lie, and came up with a picture of the galactic system which was very close to the truth.

By now we have an excellent idea of what the Galaxy is like. Seen from the edge, it would look flattened – refer back to page 95. Seen from 'above' or 'below' it would look like a rather loose spiral; this is not surprising, since many other galaxies also are spiral in form. It contains a tremendous amount of thinly-spread material, and this material blocks our view of the galactic nucleus. There are also clouds of cold hydrogen, which are not visible, but which send out long-wavelength radiation at a particular wavelength; 21 centimetres. Radio telescopes can pick up and measure this radiation, and so have been able to find out that the hydrogen clouds are distributed in the spiral arms of the Galaxy.

Other clouds of dust and gas are lit up by stars inside them, so that they either shine by reflection, are made to emit light on their own account, or do both. The most famous of these galactic or gaseous nebula is the Sword of Orion, which is a lovely sight in a small telescope, and contains a multiple star which is nick-named the Trapezium for rather obvious reasons. It is here, and in other nebulae of the same variety, that fresh stars are being created. It is not a rapid progress, but it is going on all the time.

The time has long past when astronomers could seriously think that our Galaxy is the whole universe. It is not; indeed, it is only a tiny unit in the cosmos, described in the next chapter. But, it contains objects of all kinds; and since it is rather on the senior side as galaxies go, we need not be apologetic about it. If it could be observed from a distance of, say, half a million light-years, the spiral form would be very evident. We cannot do that – at least, not yet, but certainly we have been able to find out more about it than would have seemed credible before the 1020s, when it was still believed that nothing lay beyond it.

CHAPTER VIII

The Universe

On a clear night, in the northern hemisphere, if one looks below the sharper half of the 'W' of the constellation Cassiopeia, on a line towards the bright star Beta Andromedae (see drawing), a faint milky patch is just visible to the naked eye. It is best seen, if it can be seen at all, out of the corner of the eye, since in that part vision is most effective. If one looks through a telescope at that milky patch, it still only looks like a faint foggy blur. Through good binoculars, one can see that there is really something there, and not just a dirty mark on the eyepiece.

If this dim and large shape is photographed with a long exposure with a powerful telescope, one of the most beautiful sights in the whole of the heavens is revealed. The milky patch is resolved into a great catherine-wheel of myriads of stars, with a condensed centre. It is only in fairly recent times that this glorious sight has been recognized for what it is: a whole galaxy of stars, similar to the galaxy in which we live; but a great distance away. Indeed, in looking at the Andromeda Nebula, as it is called, one has to choose one's words carefully in saying what it is that one is seeing. Strictly speaking, one is not seeing what happened a long time ago, but seeing things happening a long time ago – two million years ago, in fact. That is the length of time which it takes light to reach us from this galaxy, which can still just be seen with the naked eye. If one likes to revel in large numbers, one can say that one is viewing a collection of stars at a distance of about 10,000,000,000,000,000,000 kilometres.

We now know that our own Galaxy, of a hundred thousand million stars, is only one of a whole series of what Sir James

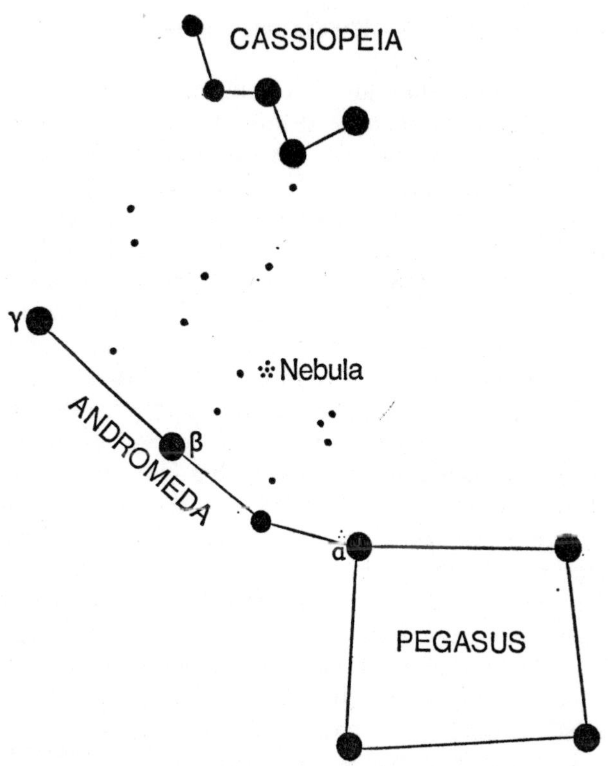

Fig. 42. Finding the Andromeda nebula

Jeans called island universes, of which we know around a thousand million. These galaxies come sometimes in groups of a
few close together, forming rough groups of various sizes. The
Andromeda 'nebula' merely happens to be the one we can see
best from Britain; there are two closer galaxies, tho Clouds of
Magellan, in the far south of the sky, which look brighter, but
which rank as companions of our Galaxy, and are much smaller
than the Andromeda system.

A good telescope will show a large number of these galaxies,
though it takes large telescopes and long exposure photographs
to reveal their shapes. Many of them consist, as do our own
and the Andromeda galaxy, of the spiral form described in the
last chapter. Some are 'broadside on' and the full spiral shape

is clear; others are seen endways on, and seem to be flat discs, with bulges in the centre.

Just as individual stars are still being born, so it is probable are galaxies. Indeed, on one theory, to which we will come presently, this is a continuous process. It is by no means impossible that, in the process of formation, the ordinary laws of physics, which we know here on earth, are not adequate to describe all that happens. It has, for instance, been suggested that, in the process of formation, quantities of anti-matter, described in an earlier chapter, are formed, eventually reacting with normal matter and releasing incredible outbursts of energy.

Certainly strange things seem to be happening in outer space. At one time, not long ago, radio astronomers were interested in sources of very powerful emissions of energy at radio wavelengths. One of these was identified with a faint blue star. When large telescopes were turned on this 'star' the observers suffered a shock. From the spectrum, it was clear that this object was not a star at all. What was more, it appeared to be at a vast distance: perhaps about one thousand five hundred million light years! Not only was the spectrum wholly different from that of a star; but incredible amounts of energy appeared to be packed into a small space. In terms of luminosity, this object, now called a quasar, or quasi-stellar object, which is much smaller than a galaxy, has a brightness equal to a hundred of our own Galaxy.

Other quasars have now been identified, and present theory cannot account for what we observe. Indeed, the whole scientific concept of physics seems to be in the melting-pot – thanks largely to the observation of one faint 'star'.

The galaxies present other problems of longer standing, and which lead us on to the borderline between science and philosophy. The first step was when measurments of the distances of the nearer galaxies were originally made. The method used was based on the relationship of certain types of variable stars between absolute magnitude and period of variation, already referred to in the last chapter. These variables were used as standard candles to calculate the distances of those galaxies close enough for us to observe the standard stars.

When a number of these determinations had been made, a strange fact emerged. Nearly all the galaxies were flying away from us at velocities which increased as the distances increased. So regular was this phenomenon that we use the velocity of the recessions of the galaxies whose distances we cannot measure directly as an indication of their distance. The appearance was that all the galaxies in the cosmos were flying away from one another. The best, though not wholly satisfactory, way of describing this is to consider a number of spots painted on the surface of a balloon. If the balloon were continuously inflated, each spot would continue to move further away from each other spot.

The method used for determining the rate of recession of the more-distant galaxies is based on the Doppler red shift, described in Chapter II. Of course, as with so much in astronomy, this principle is open to some doubt. It is not impossible to think of other causes for a reddening of light than as a result of recession. Some astronomers, indeed, do question the whole assumption upon which the expanding universe is based.

The need for caution can be illustrated by the horrid dilemma in which astronomers found themselves some years ago. The size and the age of the universe were both deduced from the recession of the galaxies, as a function of distance; these velocities, in turn, depended upon the measurement of the distances of the nearer galaxies based on the use of variable stars. Now the odd and uncomfortable fact was that the universe appeared to be younger than the earth, which is a minute part of it! The dilemma was resolved when it was found that there had been a confusion between two classes of variable stars. When this was cleared up, the age and size of the universe was doubled overnight. It may be, though it seems unlikely, that the interpretation of the red shift, and indeed the whole idea of an expanding universe, may one day be jettisoned. In that case, all our ideas about cosmology would have to be revised, and, maybe, perhaps even some of the theories about quasars, since their apparent distances are based upon their red shifts.

It is, however, currently assumed that the red shifts do represent real recession velocities, and a query at once raises its head. If the galaxies are truly receding from one another, we can project their positions back in time, when it is clear that they must have been closer together. Indeed, if we go back far enough, and for a calculable time, all the galaxies must have been bunched up together in a sort of primeval atom.

If this were truly so, we are faced with a difficulty. One can imagine a huge lump of primeval matter in a state of equilibrium; but, in that case, it would have required external interference to account for the disequilibrium which set the galaxies flying apart. When there was a lecture on the subject at the Royal Astronomical Society a year or two ago, the evening papers came out with screaming headlines saying that astronomers had proved the existence of God! If, on the other hand, the original conglomeration were not in equilibrium, we may ask in what state it was the 'day' before the universe began. There is no agreed or satisfactory answer.

Postulating cosmic theories has become something of a fashionable exercise. The subject is too deep for much examination here; but there have been a couple of theories put forward which would resolve the dilemma. The most enterprising comes from work by Bondi and Gold interpreted by that prolific source of novel concepts: Fred Hoyle. This is the theory known as the steady-state universe. Hoyle postulates that the universe is infinite and becoming more infinite all the time. As the galaxies draw apart, so matter is spontaneously created in the interstices between them, slowly forming into new galaxies to take the place of those which have receded. On this reckoning there are galaxies infinitely beyond our view, receding from us at speeds well over that of light, so that, even if the distances did not make it impossible, we could never see them.

The theory is an attractive one. True, it involves the miracle of continuously and spontaneously created matter; but almost any cosmic theory involves us in some form of incomprehensible conception. On the face of it, this theory ought to be capable of testing. If we take the alternative theory, known as the big bang theory, it becomes apparent that there ought to be an apparent bunching of galaxies towards the visible limits of the universe. This is because what we see in remote spaces is not galaxies as they are, but galaxies as they were thousands of million of years ago, when the light from them started its long journey towards us. This means that the more-remote galaxies are being seen at an earlier stage of development, when they were closer together than they are presumed to be now. The nearer ones, whose light has taken less time to reach us, show something more like their real distances apart. On the steady state theory, the galaxies would, on average, be the same distance apart, with an even distribution.

At the meeting of the Royal Astronomical Society already referred to, the distinguished radio astronomer Professor Ryle announced that he had discovered just such a bunching, and ended his lecture with the words: 'I therefore submit that the steady-state theory cannot stand.' It was a dramatic moment; but, as might have been expected, the proponents of the steady-state theory fought back, and claim that, by modification of their theory, the known facts can be accommodated. There has been much discussion since then, and it cannot be said that the matter is settled.

The other candidate for explaining the origins of the universe is, at first sight, extremely attractive. It might be called the pulsating universe. This theory would allow a finite universe, which alternately expands and contracts. It would suppose that the vast recession velocities of the more-distant galaxies are gradually slowing down until, at last, they would have stopped receding at all and would begin to contract. The contraction would continue until all the matter in the universe was congregated into a small space, when reactions would trigger off a fresh explosion, and the whole mass would again start to expand. The process would be repeated over and over again, so that there would be no beginning and no end.

Though this theory is the one which most nearly coincides with common sense and involves no ideas which we cannot easily grasp, it suffers from the defect that, so far, no one has been able to suggest any mechanism which would significantly slow down the recession, let alone start a contraction. Very recently, there has been some evidence that there may be more matter in the universe than had previously been supposed. If there were enough matter, this might provide a sufficiently strong overall gravitational field to supply the necessary force to slow down, and eventually stop, the general recession. However, for the time being, the whole question remains open. What form of cosmology anyone adopts is almost as much a matter of personal philosophical taste as of hard evidence. This is one more fascinating question which the next few years may solve in some dramatic and unexpected way.

The size of the universe is another open question – unless one accepts the view that it is infinite in size and mass. The furthest galaxies that we can see are over five thousand million light years away. We are seeing them as they were before our world was born. But, if we accept the red shift as a measure of the quasars' distances, some of them are more than seven thousand million light years away, and receding at over ninety per cent of the speed of light.

Even if it is much larger than the distances of the quasars suggest, we could never know by direct observation, because neither light nor radio signals would reach us from objects receding from us at a speed equal to that of light. One theory suggests that the universe is curved, because space itself is curved. The universe, therefore, would be finite but boundless. It is, in one way, a satisfying answer to an otherwise incomprehensible problem; but it is not easy to visualize. The nearest we can get to it is to think of two-dimensional beings, living on the surface of a sphere, and faced by the same difficulty. We can see just how their query would be resolved. To resolve our own, we merely have to conceive that we are living in an extra dimension, which our senses are not capable of perceiving.

The future – and probably the immediate future – may see great progress in finding answers to many outstanding questions. We are not even certain why the galaxies take the form they do. Many of them, like our own and the Andromeda galaxy, are in the familiar spiral form, slowly rotating about their centres. Our own sun takes about 225 million years to make one revolution round the centre of the galaxy. Other galaxies take the form of what is called a barred spiral, a rather odd shape. Still others are elliptical, and without arms. We have no idea, at the moment, what causes the galaxies to take up the varying patterns which have been observed.

The milky patch which is at the limit of visual observation may not be much of a sight to compare with the glory of Saturn's rings, or the ever-changing pattern of Jupiter's satellites and belts; but it turns out to be, in its way, the most exciting object in the sky. From one problem it leads to another, and the end is by no means in sight. Not only astronomy but physics itself is in the melting-pot. Doubts are now beginning to creep in as to whether we may not have to abandon the hallowed and cherished picture of the atom itself. It may be that our fundamental ideas about the smallest of all entities known to us may be revolutionized by our study of the vast and trackless universe.

CHAPTER IX

Telescopes and Mountings

The astronomical telescope is, in its way, a rather surprising instrument. In the first place it is, basically, astonishingly simple. The largest telescope in the world, at Mount Palomar, is fundamentally no more than a gigantic shaving mirror. In the second place, the two parts of the telescope, the mirror or lens and the ironmongery can be the queerest mixture of accuracy and crudity. Naturally, the very large professional instruments are built to very fine tolerances; but many very useful amateur telescopes are home-made, but still extremely effective. My own 12″ has a mirror accurate to be better than a tenth of the wavelength of sodium light, a main drive from an electronic device of good accuracy, and a slow motion controlled by an old windscreen motor and some Meccano parts!

The original form of astronomical telescope consisted of two convex lenses, fastened into a tube to keep them the right distance apart. A very easy way to grasp the principle is to buy or borrow a spectacle lens used by a long-sighted person. This will merely be a convex lens of long focal length. Let us repeat the experiment partially described in Chapter II. If one holds up such a lens in a room in front of a window, but not too close, and places a sheet of thin paper behind it, an image of the window, upside down, appears on the paper if one moves it to and fro. There will be a point at which the image is clearest, and this is when it is in focus. If one then takes a reading glass, which is only a convex lens of shorter focal length than the spectacle lens (i.e. it is more sharply curved), and holds it behind the paper, it will magnify the image on the paper. Actually the paper is serving no useful purpose. If it is removed, the enlarged image is still visible. This is an astronomical telescope in very simple form. The two lenses can be held in any suitable kind of clamp, which will allow the relative distances to be varied, and the moon can be viewed through the contraption well enough to see the craters. For this purpose, the lenses will have to be moved further apart, to bring the image of the moon into focus. In the case of a terrestial telescope, there is an extra set of lenses to turn the image the right way up. Since, however, some light is lost in every passage through a lens, and since an astronomical object can be observed equally well either way up, the rectifying lenses are omitted in an astronomical telescope. It is for this reason that pictures of the moon, or indeed maps, are generally printed the 'wrong' way up and the 'wrong' way round. The same is true of other astronomical pictures. On the other hand, one should note that to someone in the southern hemisphere the moon is seen what, from the northern hemisphere, appears to be the wrong way up.

The principle of the convex lens is quite simple. One can consider it as an infinite series of truncated prisms, which bend light, as described in Chapter II. If we make a drawing of the light passing from an illuminated object through a convex lens, we can trace how light spreading out from one point appears behind the lens concentrated again at a point opposite to that of the source. The drawing on page 46 should make this clear. Experiment or theory will show that the real image made by a convex lens is further away from the lens, and larger in size the *less* the curvature of the lens. On the other hand, when viewed through a convex lens from close to, the *greater* the curvature of the lens, the greater is the magnification. It follows that for maximum total enlargement one needs a very slightly curved lens at the front, called the objective lens, and a sharply curved lens, called the eyepiece, at the back. The limit to the useful magnification with any particular telescope is the amount of light it gathers. The amount collected depends upon the diameter of the objective, but the more one magnifies the primary image, the dimmer it appears. Clearly then, the maximum useful magnification which can be employed depends upon the actual size of the objective.

The kind of crude telescope we have been describing has serious limitations which were recognized by the early makers.

If one tries such a telescope on a bright object like the moon or a star, there are a series of colour fringes around the image, which go a long way to ruin it. This is because light of different colours is refracted by different amounts by any one kind of glass. The result is that, since white light consists of all the colours of the rainbow, the different constituent colours are brought to a focus at different places. The blue light will be in focus when the red is substantially out of focus and vice versa.

A way out of this difficulty is to use a compound lens of different glasses, which have opposite differential effects. All good refracting telescopes have such compound lenses. Many a useful refractor has been ruined by a novice taking out the objective to clean it, and putting it back the wrong way round. Some lucky people have secured bargains by buying a telescope thus treated, and then putting the objective back properly!

To begin with, however, before such methods of dealing with the problem had been discovered, Newton made a reflecting telescope which used a mirror in place of an objective lens. Simple experiment will show that a concave mirror has the same property as a convex lens of producing a real inverted image. Nor does such an image, made by a mirror, have any colour fringes. It has, however, one substantial drawback. If one borrows a shaving mirror of the kind which magnifies the image of the face, and tries to produce a real image on a bit of paper, as we did with a lens, the trouble is at once apparent. The paper blocks the light out from the mirror, since it needs to be held in front of the mirror.

Newton overcame this difficulty most ingeniously. He placed a small flat mirror in front of the larger one, and at forty-five degrees to it, commonly called the flat. This catches the converging beam from the main mirror, and diverts it at right angles to one side, where an eyepiece can be placed. The drawing will help to make the system plain. (Fig 43)

Although the reflecting telescope was devised to overcome a particular difficulty which can be removed in another way, reflecting telescopes have come into very general use in astronomy. All the largest telescopes in the world are reflectors. If for no other reason, it is impossible to make very large lenses, which would be incapable of sustaining their own weight. Moreover, the thicker part of the lens would be so great that a large

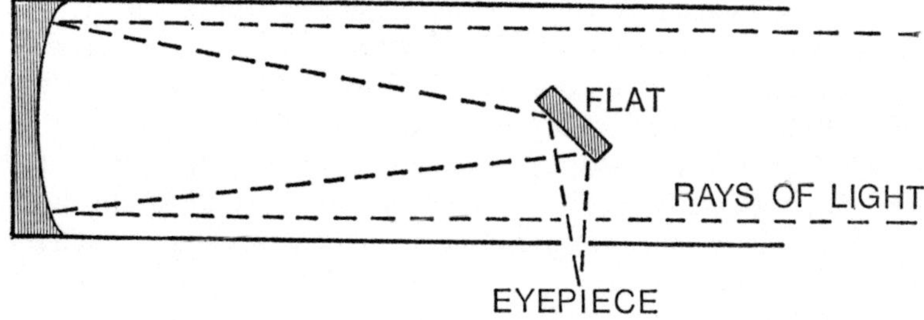

Fig. 43. The principle of the reflecting telescope

amount of light would be lost. For the amateur, there is another advantage. Many amateur telescopes, including some excellent ones, are completely home-made. It is possible to grind a mirror, but not a lens, for oneself. Further still, a refracting telescope tends to be much longer than a reflector of the same size. A twelve inch reflector may only be six feet or so long. A refractor of six inches needs a big dome to house it, and is relatively unwieldy.

Grinding one's own mirror is a lengthy and delicate business, especially as the curve has to be paraboloid rather than spherical. However, it is not impossible for the neat-fingered and patient, and a reflector of substantial size and good accuracy can be made at home at little more than the cost of buying the blank or glass or Perspex from which the mirror is to be ground. Once this has been done, an aluminium film has to be deposited, since it is this aluminium film, rather than the glass or Perspex, which does the reflecting. For the few people who keep telescopes very close to the sea, as indeed I do, there is a fresh problem. All coated mirrors need to be re-aluminized at long intervals, probably of several years. The sea air, however, can ruin the reflecting capacity of an aluminium film in a matter of a few weeks. Cleaning helps, but is only partly effective. A silicone coat over the aluminium is another way to improve the life of the mirror coating, but it will not counter really heavy tarnishing. The secret is to use rhodium in place of aluminium. There is a drawback, in that rhodium reflects less light than aluminium. On the other hand, it is better to have a fair image which endures than a better one which only lasts a few weeks. Rhodium can be washed clean again and again with little trouble. It costs a trifle more to have in the first place, though neither costs more than two or three pounds, according to the size of the mirror. Unfortunately, rhodium coating is not to be tried and discarded; once it has been put on, it cannot be removed at all easily.

A reflector has one disadvantage compared with a refractor: the flat, though it blocks out only a tiny fraction of the light reaching the main mirror, does cause a small amount of diffraction of the rays passing near it, as does the array mounted to hold it in place, and called the spider. A reflector, therefore, has a slightly less clear image than a refractor, size for size. By and large, one reckons a four-inch refractor as about the equivalent of a six-inch reflector for definition, though the larger reflector will have more light gathering power, provided that the aluminium coating is fresh and clean.

Which kind of telescope one will choose depends upon the use to which it is to be put. When looking at a distant and faint object, the chief concern is the diameter of the mirror or objective, since the amount of light grasp increases as the square of the diameter. For near and bright objects, the great concern is with accuracy of the definition.

The ordinary Newtonian reflector, which has been described, is the most common type of reflector; but there are variations. In other types of reflector, the flat is replaced by either a convex or concave circular secondary mirror, set parallel to the main mirror, the reflected light beam from the secondary is passed through a circular and a small hole in the middle of the main mirror, and viewed through the back with an eyepiece. One result is that the effective focal length of the telescope is nearly doubled, as is the size of the primary real image. A much shorter and handier telescope can thus be constructed though it has also certain drawbacks. It needs, for instance, to be mounted on a much higher stand, so that the eye can be got to the back, when the telescope is pointing towards the zenith.

A reflecting telescope, then, consists of only three effective parts: a main mirror, a flat to throw the beam sideways, and an eyepiece through which to view an enlarged image. To meet these requirements, the ironmongery holding the optical parts can be very rudimentary indeed. One possible method is merely to mount the mirror on one end of a simple spar, with the flat

on a projecting arm at the other end. The eyepiece can be fixed
to another arm at the forward end.

The more usual method is to mount mirror, flat and eyepiece
in a tube, which may be open work, or a solid tube. The first
version of my own 12" made by Calver in 1889, had a closed
tube nine feet long. With so long a tube, the primary image was
of fair size, but the inconvenience of viewing was marked.
When pointed towards the zenith, it was, naturally, necessary
to raise the eye nine feet above the ground, which involved a
longish ladder and some tricky gymnastics to get one's eye to
the eyepiece and keep it in position as the star observed moved
round. My later version has a shorter focal length, with a tube
only just over six feet long and made of open lattice-work. To
help with keeping the eyepiece in a suitable position, the end of
the tube in which the flat and eyepiece are mounted can be
swivelled round, bringing the eyepiece to some spot suitable
for the height of the observer. An alternative method achieving
a similar result is to mount the tube so that the whole of it can
be rotated.

With a reflecting telescope, it is important that the flat, mirror
and eyepiece are correctly aligned with one another. To assist
this object the 'cell' or holder in which the main mirror is
mounted is generally fitted with adjusting screws, enabling the
mirror itself to be moved slightly in any desired direction. These
adjusting screws normally work on floating three-point suspen-
sions for the mirror, thus avoiding pressure points on the back
of the mirror.

This kind of precaution is very necessary, and the amateur
who first acquires a reflecting telescope must start by learning
respect for his mirror. The ironmongery of a telescope can be
robust, but the mirror is tender beyond belief. The slightest
uneven pressure on the mirror will distort the image. When it
needs to be cleaned, it must be done with great delicacy. The
surface may only be touched lightly with soft cotton wool. Any
pressure may spoil the aluminium film, or even distort the mirror
itself. When the telescope is not in use, the mirrors are normally covered with closely fitting lids, and the telescope itself
housed in a movable shed or dome. When prepared for use,
time must be allowed for the temperatures to even out; a slight
variation between the different parts of the mirror will distort

the image. Even the warmth of a hand, held in the tube, will set up a thermal current of a disturbing nature. On no account allow visitors or children to touch either of the mirrors. A fingerprint, unless instantly and delicately removed, will leave a permanent mark, and heavy pressure may permanently distort the mirror. Because of the thermal disturbances caused by slight inequalities of temperature, it is impossible to heat an amateur observatory. The observer must wear enough clothing to stand up to a freezing January night! It is possible for an expert to make a heated observatory by having the telescope itself in the open, and bringing the rays of light in through a channel in the axis; but great skill and accuracy are needed to do all this satisfactorily. I know of only two such amateur telescopes.

The flat, at the other end of the telescope, needs equal care in mounting and provision for adjusting its distance from the main mirror and for its angle to the mirror and eyepiece. It is generally mounted in a holder suspended from the tube by three thin sheets of metal parallel to the tube. The cell, or holder of the flat, can be held at the forward end by a rod in a tube, equipped with a locking screw. By pushing the rod forward or backward the distance from the main mirror can be adjusted. The whole of the cell and rod are also on a ball-and-socket joint, and the angle can be adjusted by loosening the locking nut on this and moving slightly until the angle is just right.

The eyepiece is, in effect, merely a sort of microscope, through which the primary image is viewed. Ignorant laymen are fond of asking what the magnification of the telescope is. The answer is that it entirely depends upon what power of eyepiece is used, and one needs to have a selection. It is a beginner's temptation both to use too high a power, and then not to be too truthful about the magnification actually employed. For a twelve inch, one needs a low power eyepiece of about 40/50 magnifications of the primary image. The power is usually written as x50, which indicates fifty magnifications. A mid-power of about x100 is useful, with a maximum power of x200/300. Over that, the image will be so faint and blurred as to be pretty useless.

Most eyepieces consist of compound lenses, but one may have a single lens, as with a Tolles. One very useful kind is the Barlow, which is an eyepiece of adjustable power. If one takes any stan-

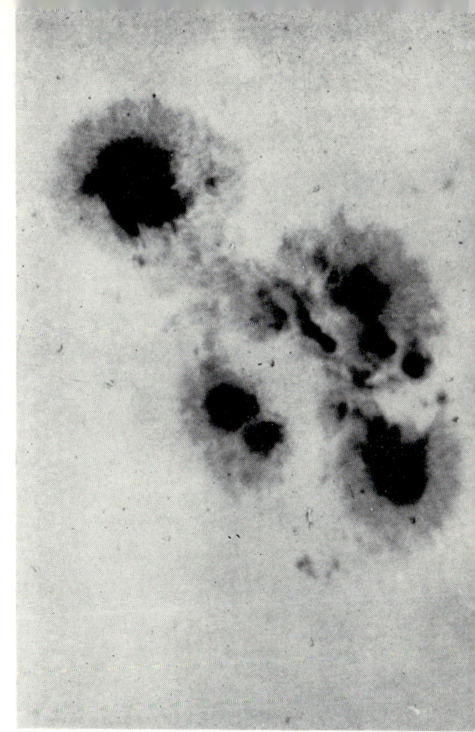

Left The eclipse of the sun in March, 1970, photographed in Mexico and transmitted by television, from which this photograph was taken. *Right* Sunspots, photographed by W. H. Baxter with 4″ refractor. *Below* Bennett's Comet, photographed by F. J. Acfield in April, 1970

Top Perseid meteor, photographed by H. B. Ridley. Note the star-trails, due to the long-time exposure. *Middle* Andromeda Galaxy, photographed with the Palomar 200″ telescope. *Bottom* Spiral galaxy in Leo, photographed with the Palomar 200″ telescope

dard eyepiece and places a simple concave lens so far back that it comes behind the primary image, the primary image is increased in size, and projected slightly forwards. By having such a concave lens mounted in a way which permits its distance from the eyepiece to be adjusted, the magnification of the eyepiece can be varied by about four or five times.

The position of the eyepiece needs to be easily and accurately varied, according to which eyepiece is being used. One cheap method is to have a tube, into which the eyepiece screws on a standard thread which itself slides easily in a slightly larger tube attached to the telescope, the movable tube being known as the draw-tube. Since the focusing of a good telescope is a matter for very great accuracy this simple focusing arrangement is not really good enough. The normal practice is to have a tube, attached to the body of the telescope, which can be moved in or out with a rack on the tube, operated by a pinion on a large focusing knob. The draw-tube fits into this, with the eyepiece screwed into the end. The crude focusing is done by pulling the draw-tube in or out, and, when the image begins to come into focus, the final adjustment is made by using the focusing knob.

All astronomical telescopes need mounting. The field of view is far too small to permit of hand holding. The mounting, moreover, needs to be very steady and the telescope on it easily and smoothly moved. It is one of the drawbacks of small refractors that they are seldom steadily mounted, which makes them useless for observation. One gets the star, or whatever it is, in the field of viewing, and finds that lining the telescope has set the image wobbling intolerably. By the time the image has steadied down the star has moved across the field and passed out of view!

There are two commonly used types of mounting having in common that they are rock-steady and the telescope moves on them easily and without vibration. The cheaper and simpler mounting is the Altitude-azimuth (Alt-Az), on an upright pillar. The telescope can be moved along the horizontal plane (Azimuth) on one pivot, and up and down (Altitude) on another, so that, between the two, it can be directed at any point of the heavens. The drawback to this simple system, and it is a considerable drawback, is that constant alterations are necessary in both

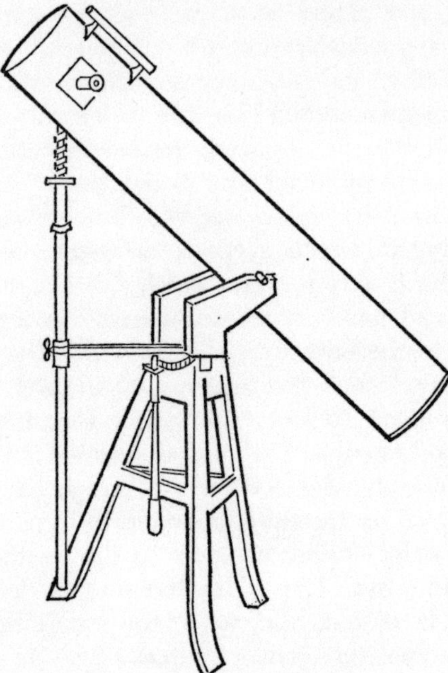

Fig. 44. An alt-az stand

directions to keep a celestial object within the field. It must be remembered that a star moves not only from east to west, but also from high up to low down and vice versa.

With any kind of mounting, but especially with an alt-az, there needs to be some mechanism for making small adjustments, in order to follow the course of the star. The first crude movement, picking up the star and getting it into the field of vision, is done with the telescope unclamped; but, once it has been found, direct hand movements are too crude with any considerable magnification. With a powerful eyepiece, the field of view may be as little as five or six minutes of arc. Even with a very low power, it is probably only about thirty minutes, which is just large enough for the moon to occupy the whole field. In the higher power, the object will move right across the field in a matter of a few seconds, and if one attempts to follow it by pushing the telescope by hand there will be no joy at all.

Two hand-made adjustments are required, one to move the

telescope by very small amounts if azimuth, and the other to move it in even-lesser amounts in altitude. A small worm and wheel attached to the mounting and the telescope can be employed to adjust in altitude, and another can move the pivot in azimuth. This necessitates using both hands to steer the telescope, and leaves no hand free to make notes, nor even to adjust the focus. Moreover, it is not possible to arrange for a motor drive with an alt-az stand, and a motor drive is a necessity for some purposes, such as photography.

Any form of stand with a useful telescope requires means of first finding the object of viewing. As we have noticed, even a very low power only gives half a degree of field, and there are a hundred thousand such fields to choose from if there is a clear view of the whole sky. Just to pick up a star by pointing the telescope in the right direction can be baffling. The solution, universally used, is to mount a much smaller telescope, known as the sighting telescope, on the tube of the larger one. The sighting telescope has crossed wires in the middle of the field, and is so adjusted that when the object to be viewed is on the cross wires it will be in the field of the main telescope. With a telescope of good power, even the sighting telescope may have a smallish field, if it is to be able to pick out dim objects. On my own twelve inch, I have a two-inch refractor for a sighting telescope, which has a two degree field. For quickness of working, I have a second sighting telescope of even lower power and wider field, a five degree one.

For a useful telescope, capable of a wide range of uses, the mounting is what is called an equatorial mounting. In this case the main axis of the telescope is sloped, so as to be at the same angle from the ground as the latitude of the observatory. At the pole, the axis would be vertical, at the equator it would be horizontal. In the south of England it will be at about fifty degrees. The angle is measured from the ground in the direction of the north and the whole telescope is aligned to point due south.

When the axis is so arranged, an enormous benefit occurs. Once the telescope is pointed at any star, a mere movement of the instrument on the main axis follows the star round, without any alteration in the altitude. This means, amongst other things, that the telescope can be mechanically driven. If the speed is

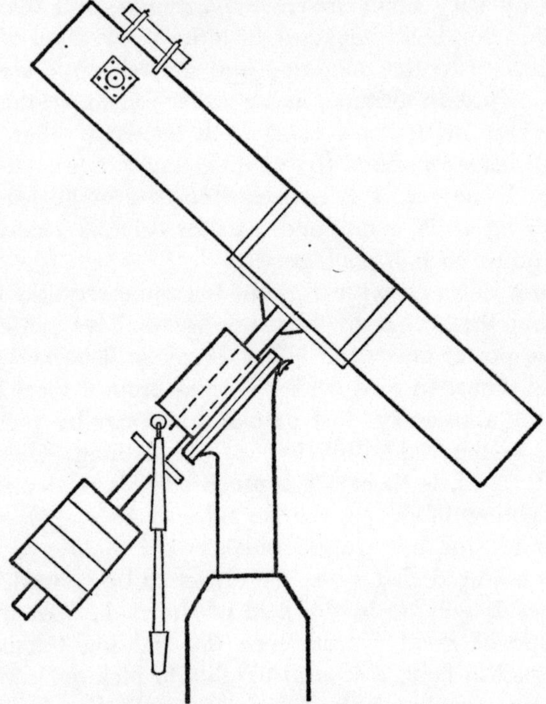

Fig. 45. An equatorial stand

accurately timed at one revolution in a sidereal day, the star will stay fixed in the field without any adjustment at all. Apart from convenience, this is essential for taking photographs, other than the very few which can be taken without a time exposure.

Another advantage of the equatorial mount is that, with the aid of graduated circles, the telescope can be pointed in the right direction without looking through the eyepiece at all. It is only fair to add that most amateur telescopes do not have accurate enough circles to pick up an object in high power without some searching. This can generally be done by clamping the telescope in declination (that is, the up and down movement), and searching slightly in right ascension (that is to say sideways).

The declination circle is attached to the stand of the telescope and has a pointer on the tube of the telescope itself which moves as the instrument is pointed towards the pole or away from it. It is graduated in degrees and fractions of a degree, and so fixed that the reading is +90 when the telescope is pointed at the North Pole, and zero when it is on the celestial equator. Below the equator, towards the horizon, the scale reads from zero upwards in degrees minus. It can be seen that if the telescope is set on the declination circle to the declination of a given star, and then turned on its axis right round, the telescope will, at some point, point to the star. Alternatively, if the instrument is left still, the movement of the celestial sphere will, at some stage, bring the star into the field of the telescope.

To avoid the need to sweep far when the declination is set, there is also a circle which marks out the right ascension in hours and minutes. One circle is fixed to the main axis of the telescope, and moving round with it, with another fixed circle above it and fastened to the stand. The circles are so graduated that when the telescope is pointed due south (in the northern hemisphere) the reading is zero. For each fifteen degrees the axis is moved, one hour is registered on the circle. Another way to put it is that, if the drive is working, for each (sidereal) hour that passes one hour will be registered on the circle. It is child's play to look up the declination of any celestial object, and set the declination circle properly. It is not quite so easy, but still moderately so, to work out the Right Ascension (written R.A.) of any star at any given *local* time, and set the circle accordingly. In theory, and if one's arithmetic and circles are accurate, one can look through the eyepiece and see the required object. In practice, with amateur telescopes, it is generally possible to set the declination and R.A., clamp the telescope in declination and sweep a little way from side to side by hand.

There are some sets of circumstances in which, having got the telescope properly set and accurately driving, there is occasion to make small adjustments. If, for instance, one is looking at the moon in high power, only a small portion of it will be in the field. To see some other part, adjustment in both R.A. and dec. will probably be needed. Even if this is not so, the moon, as well as partaking of the normal movement of the celestial sphere, is moving with its proper motion in both dec. and R.A. To keep it in view, it will be necessary to move in both directions from time to time. To undo the clamp, and adjust by hand, would be time consuming and not very easy to do in high power without losing the moon.

With an equatorily mounted telescope, the adjustment in dec. can be made as described with the alt-az by a screw; but the R.A. alterations are not quite so easy, since the drive is rigidly fixed to the axis. The difficulty is overcome by not driving directly from the motor. The worm of the motor can drive a circle, or a quadrant, not fixed to the axis but driving another circle, or quadrant, through a worm. By driving this second worm, ether with a cable or a secondary motor, the telescope can be moved slightly in R.A. without detaching the drive motor. On my own telescope, I use a small electric motor, with a reversing switch, that can be held in the hand, to turn the secondary worm. In this way, and without moving the eye from the eyepiece, one can adjust both R.A. and dec. by small amounts. Naturally, at some critical moment, one finds oneself at the end of one of the worms. It is, therefore, as well before starting to observe, to make sure that both worms are near their centres.

What has been said about reflecting telescopes can also be applied to refractors, though it is less usual to find amateurs with efficiently mounted and driven refractors. If for no other reasons, refractors, if professionally made, are expensive, and cannot be made at all by an amateur. Many beginners start with a small refractor, of perhaps 3″ aperture, and not well mounted. The result is usually an image which wobbles like a jellyfish, and has a genius for disappearing altogether just as one is beginning to see something. If one is to start with a 3″ refractor, it is worth a good deal of pains and cash to make sure that it is mounted firmly and efficiently. This normally costs as much as buying a 6″ reflector, which, for most purposes, is superior.

To buy telescopes, new, even small reflectors, can be costly. There are generally secondhand ones to be found at far lower cost. Still better, if one has neat hands, is to make one's own, even if one has to buy the optics. One piece of advice, however, is indispensable. On no account should a beginner buy a telescope, either new or secondhand, without advice from some more-experienced observer. Also, it is generally a waste of money to buy a very small telescope. It may be all right for a child to start with a 2″ refractor or a 4″ reflector, but no useful work can be done with either. Even for real interest, a properly mounted 3″ refractor or a 6″ reflector is the smallest that is of much use.

CHAPTER X

Equipment and Observing

The moment that one graduates to a telescope of moderate performance and size, a problem of housing presents itself. Such an instrument cannot be picked up and moved about. In its original form my own 12″ weighed the best part of a ton! It, however, was built when brass and cast iron, as well as lead for counterweights, were cheap. The first need is for a firm foundation, and it is difficult to improve on a concrete base set into the ground. It is possible, once the instrument is in place, to cover it up with plastic, or some other form of sheeting; but the practice is not to be recommended. It is a nuisance to put on and remove, and the wind, at some stage, will certainly tear it or rip it off.

The best, but most expensive, solution is the professional one of mounting the telescope inside a dome. In this type of observatory the walls are generally fixed, and the roof, which is circular and with a panel which can be opened, revolves on runners. In the best kind, one points the telescope through the opened panel, and the roof is driven at the same speed as the telescope itself, so that the opening follows round with the instrument. Slightly more cheaply, the opening can be rather larger, and the roof pushed round by hand from time to time.

Unless one is very good with one's hands, or has an adept and kindly friend, a dome is prohibitively expensive, and apt to cost a good deal more than the telescope itself. An alternative, and the most widely adopted one, is to have a run-off shed. This can be made in either of two forms. My own is one I built myself, though it would be hard to find anyone clumsier with his hands. It consists of hardboard on a wooden frame, with a glass fibre roof, pitched to let the rain run off. One end of the shed may be a door which can be lifted off altogether, and the shed merely pushed away on rails, which consist of 1½″ angle iron on concrete tracks. It is only a matter of half a minute to lift off the door and push the shed away. When in place over the telescope, it provides complete shelter. An alternative, as Patrick Moore has over his 12½″, is a shed without a door, but built in two halves. Each half is mounted on rails, and can be pushed away in opposite directions. There is not a lot to choose between the two methods.

If one is making one's own telescope, there are two main alternatives for mounting it. One common way is with a fork mounting. This consists of a fork, with the telescope mounted between the prongs, which can swivel on a pivot. The movements in dec. are made by moving the tube up and down in the fork. The R.A. by rotating the fork itself. An old car axle can form a useful basis for the pivot.

The alternative is to fix the telescope on one side of an arm on the end of a pivoted and substantial rod, held in a rigid frame. In such a mounting, heavy counter-weights have to be fixed on the opposite side of the arm, to balance the weight of the telescope. The photograph (opposite page 32) will help to explain the system.

There is scope for a good deal of ingenuity in collecting one's observatory accessories. The first need is for a pair of steps, in order to reach the eyepiece when the instrument is pointed towards the zenith. The steps need to be good and sturdy, both to withstand a lot of exposure, and to make sure that the astronomer does not come to a sudden end because the steps have overbalanced or collapsed under him.

A most desirable adjunct is an observing box, to hold one's eyepieces and other odds and ends of equipment, including a torch to compete with failures in such parts as the drive. It should however be remembered that the torch should only give a weak light. It takes about twenty minutes for one's eyes to become adjusted to the darkness, and it is most provoking if one has to start again at scratch because one has blinded oneself with a brilliant torch. For a similar reason, it is useful to procure a pencil with a built in, but very weak, torch bulb. Often one wishes to draw at the eyepiece something like a crater on the moon or the belts of Jupiter, and a combined pencil and torch is extremely useful.

Another accessory, very useful for some purposes, is a filar micrometer. This consists of an eyepiece fitted with two parallel hair lines, one of which can be moved by turning a knob at the side. When measuring a distance – for instance, the separation between the two components of a double star – it is possible to turn the knob and rotate the eyepiece until one line rests exactly on each star, and then read off the angular distance on the micrometer. It has other uses; for instance, ascertaining the phase of Venus.

Yet another useful possession, if one is going to do advanced work, is a spectroscopic eyepiece. This may be prismatic, but is more generally a diffraction grating, which has the same property as a prism of splitting up the light of the star into its component colours. Spectroscopy, however, is a highly technical subject, and only of marginal interest to amateurs.

More useful is a set of colour filters to fit on the outside of the eyepiece. Red, yellow and blue are the ones usually chosen. Also, for some purposes, a polarizing filter, or a pair of them, come in handy. If two are used together, one needs to have them fitted so that one can be rotated over the other. Used singly, a polarizing filter is chiefly of use in cutting out glare from over-bright images. The two together enable one to regulate the amount of light emerging, by rotating one above the other. When the two planes are at right angles, there is theoretically no light passing through. (In practice there is a small amount of blue light which comes through.) Ordinary colour filters have different uses. For example, the crescent of Venus will appear slightly thinner in one colour than in another, as can be verified by using a micrometer with the colour filters. Sometimes, also, craters of the moon will appear more distinct in polarized or coloured light.

Another accessory which should be owned, let alone used, only by someone who knows well what he is doing, is a sun diagonal. This is a gadget to insert between the draw-tube and the eyepiece, which reflects solar light by a sheet of plain optical glass. Most of the light passes through, and only a fraction comes to the eyepiece. All the same, and even if used with a heavy filter, this is not a toy for beginners.

An important matter is the choice of eyepieces. There are whole ranges of different types, each of which has its own particular advantages and drawbacks. For the beginner, it is as well to start with only a couple, and extend one's extravagances when one has become adept. A good choice is a low power, with a field of vision of about half a degree, or a little more. This will just accommodate the whole of the moon. Its focal length will depend upon that of the telescope itself, the total magnification being a function of the focal length of the telescope, and inversely as the focal length of the eyepiece. For more detailed work, it is a good idea to get a Barlow which will give magnifications from about x80 to, say, x350. With such a pair, a whole range of viewing is possible.

Other useful accessories include a sidereal clock and a star atlas. The latter, indeed, is a must rather than merely desirable. The atlas which is very widely used is *Norton's Star Atlas*, which gives all stars down to the sixth magnitude, as well as over fifty pages of useful basic information. As for the clock, there is no need to go to a great deal of expense. A cheap alarm clock can be regulated to gain about four minutes a day, and will serve for most purposes if it is corrected about once a week. It is only when one comes to exact timings that extreme accuracy is required, and then the cost of a really good astronomical clock is prohibitive. All the amateur needs is a clock by which to set his R.A. circle approximately, without doing a sum each time.

Professional astronomy, unfortunately, is leaving the amateur far behind in many fields. The old enemy, lack of light, is yielding to modern electronics. It is possible to get image intensifiers, in which the light is received and multiplied electronically. Electronics, in various forms, are, indeed, rapidly becoming a major tool of astronomy. Even at the moment of writing (February 1970) a device has been announced by Edinburgh University, which will scan photographs of star fields, and, with the aid of a computer, furnish in a few minutes information which would have taken days of patient labour. Unfortunately, these devices are incredibly expensive. The amateur, however, has no need to feel despairing; there is still a great deal of work he can do, some of it of considerable use to the professionals.

When one passes from choosing and equipping the telescope to its possible uses, there is a long list, even for the tyro. Indeed one should really start with the possibilities of a pair of binoculars. Alcock, an amateur of great experience and skill, has discovered, as already mentioned, in a very brief interval, no less than four comets and three novae. Comet searching is a pastime which calls for a great deal of enthusiasm and patience. Alcock's fame was the reward of many, many hours of fruitless searching.

In addition to comets, binoculars may also be used to search for novae, though no success is likely for an observer not well acquainted with the star field he is examining. In any case, the advent of nova is rare, and most likely to be discovered by photographic comparison. A still further use for binoculars is in tracking artificial satellites.

Once equipped with a telescope of reasonable efficiency, the list of employments open even to a beginner is long. To start with the moon, there is still some work to be done with what one might call lunar geography. It will be some time yet before exploration of the surface by direct observation can be anything like complete. In the meanwhile, there are many details to be studied, and new ones discovered such as additional domes. Especially is there a great interest in reports of transient colours, which, until lately, were discounted.

Apart from surface features, some observers specialise in occultations. From time to time (a list is given in the British Astronomical Association Handbook, as well as in comparable publications in other countries), the moon passes in front of a star bright enough to be seen through a telescope in spite of the moon's brilliance. By carefully watching and the use of a stopwatch it is possible to time the occultation to a fifth of a second. Information so derived is serviceable in gaining exact information about some of the very minor irregularities of the moon's movements.

The planets are a fruitful field for the amateur, since there is still not a lot of observation of them done by the professional observatories. Venus offers a good deal of interest and entertainment. As we have noted, it is the lazy man's joy, as it is best observed in broad daylight. The drawback is that a telescope with fairly accurate circles is needed to find it. At that, it is essential that the eyepiece should be left in exact focus, or else the point be clearly marked in some way. Though even near the sun, it is a bright telescopic object, the least inaccuracy in focusing makes it difficult, if not impossible, to pick up.

By a somewhat clumsy and time-consuming process, it is just possible to find it in daylight without circles. The first thing is to find a convenient star which has exactly the same declination as Venus will have at the time when it is to be observed. Then, the previous night, one fixes the telescope on the chosen star, at a time when it has the exact R.A. as Venus will have at the time on the following day when it is to be observed. With careful sighting and accurate arithmetic, the planet can then be found. Amongst the other drawbacks of this method, especially in England, is that having worked hard and accurately at night, it is one of the immutable laws of nature that it will be cloudy when the moment comes for observing on the following day.

Among the things to look for on Venus are the shapes and changes in the cusp caps, and possible faint markings on the surface. If one is to be a regular observer of the planet, it is desirable to have colour filters and, if possible, a filar micrometer. There is also usually a group organized for a study of the period of dichotomy, which is best determined as a mean of the calculations of a large number of observers.

Mars is also a good object of study, though there are years when it is so badly placed that viewing is difficult. Among the points of interest are the changes in the surface markings, the occasional clouds and the ebb and flow of the polar caps as the seasons change.

Jupiter is a great favourite. Most of what we know about the changes and patterns of the cloud belts is due to the carefully observed, drawn and recorded work of amateurs. Since the changes are unending, there is always something fresh to be seen and noted. There is also interest in timing the motions of features in the cloud belts, as well as phenomena of the satellites.

Saturn, though so fascinating a sight in itself, offers rather fewer opportunities for profitable observation. There are occasional, though not very distinct, markings to be seen on the surface, and the movement of the satellites repays attention. The planets beyond Saturn are really of not much interest to amateurs. There is, however, a further field for study in the asteroids, as and when they are visible. It may even be possible to discover a new one.

The stars provide two fields of work for the amateur, over and beyond the search for novae. The first is in the measurement of the separation of double stars. There are very many of these objects, and the measurements given are often those from observations made long ago, and not always very accurate. A filar micrometer is a must for this branch of study, but it is a skill fairly easily learned, and offering concrete rewards.

The second, and very profitable, field for study is the regular measuring of variable stars – a special province for the amateur. A fair amount of practice is needed, since the only way, other than by highly skilled photography, to make good assessments is by selecting comparison stars of known magnitude, and in the same field as the variable. When the observer is adequately skilled, there is also a possibility, not too remote, that he will discover new variables.

The last few years have opened up a new branch of observational astronomy to the amateur, and this is the study of artificial satellites. Mainly this is done with the naked eye, or with binoculars; but if telescopically observed – a difficult feat – the value of the observation is greatly enhanced. The basic principle is to time with great accuracy the exact position of the satellite against the background stars. From many such observations the exact track of the satellite can be worked out. It is provoking enough sometimes to do this work with binoculars, let alone a telescope. First one must process the given data, to know in what position and at what time the satellite can be looked for. If a telescopic sighting is intended, the work must be done with great care and in great detail, since the given data will differ even for observers comparatively close together. After all the labour of calculation, the whole often proves fruitless because of cloud when the moment comes.

Reference has been made to searching for comets with binoculars. The same work can also be done telescopically. Indeed, the comet will be visible in a telescope long before it can be picked up with binoculars. The drawback is that the field of view of the telescope is so much smaller that the chances of finding a new comet are remote indeed. A telescope is, however, needed when a comet has been discovered in order to study and

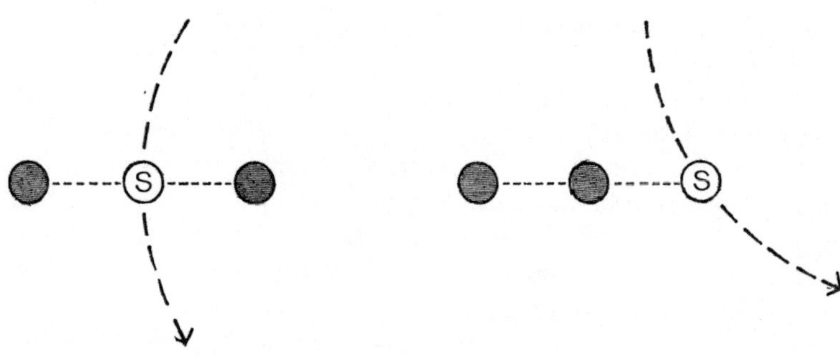

Fig. 46. S is an artificial satellite. The two figures show possible ways of recording its position at a given moment in respect of a couple of stars near which it happens to be passing

follow it carefully. In relation to this sort of observing, an amateur can do worse than to make regular studies of one particular star-field. If a suitable area is chosen, and the field looked at until it has become perfectly familiar, there are quite a lot of possibilities in the way of discoveries – novae, comets, variables and asteroids.

The study of the sun is a rather specialized occupation. If it is to be a chosen speciality, one needs a specially equipped refractor, which must only be contemplated by a fully-experienced observer, who has worked for some time with a more-qualified practitioner. For the beginner, there is quite a lot of interest, *provided that one sticks rigidly to the rule that, in no circumstances, must one look at the sun directly.* The right method is to project an image on to a screen from a refractor. If one puts in a lowish power eyepiece, and then moves a sheet of white paper behind the telescope, which has been directed at the sun by approximation, it is possible to get the telescope properly lined by moving it slightly in different directions until a bright beam emerges. By varying the distance at which the paper is held, a place will be found where there is a bright and sharp image of the sun projected upon the paper. The image will be sharp enough to perceive and study any sunspots which may be there at the time. A better alternative, once the principle has been grasped, is to fix a screen on the end of the telescope, at about the right distance; the final focusing can then be made by movement of the eyepiece. If the telescope is driven, the image on the screen can be kept steady for long periods, thus enabling the sunspots to be studied thoroughly.

There is one other field open to the amateur. Perhaps it ought to have been mentioned first, since no equipment is needed at all, except the eye of the observer, or perhaps a simple camera. This is the study of meteors. A good deal of interest centres on these more or less random entities. At the time when a shower is due, it is worth while taking a little trouble. Once more, naturally, experience is a help. There is only a short time – a matter of seconds – in which to make a useful observation. Sometimes, indeed, there may be more than one meteor visible at the same time. The object of watching is to be able to describe the apparent motion of the meteor against the stellar background – and observations will vary with the position of the observer. One also needs to study and report on many details of the meteor: its brightness, what sort of a trail it makes, its colour, and so on.

In this short list no attempt has been made to describe in detail either the methods or the full scope of the useful and interesting observations which can be made. A whole book would be needed to do that. The real point is that astronomy is a co-operative undertaking. Even before one buys one's first modest telescope, the budding astronomer would do well to join one of the associations of amateur astronomers. In Britain, the leading national society is the British Astronomical Association. In the United States and most other countries, there are similar organizations. Membership of such a society is of help, not only because it offers a clearing house for observations and issues helpful information, but also because the best way to learn the subject is to join in the lectures and discussions which are frequently held. Advice is always available, and most such societies have separate sections, any or several of which can be joined, for co-ordinating work in most of the branches of observation which have been briefly listed above.

CHAPTER XI

Photography and Radio
Astronomy

Photography is a separate branch of the observing technique. To do much with it requires fairly sophisticated equipment, though some excellent work is done by those who have made their own telescopes and cameras. Without a driven telescope, there is little scope in photography. One may, indeed, start by pointing an ordinary camera approximately at the pole, and leaving it open for the odd hour on a clear night. The result is a series of concentric part circles. Such a picture at least helps one to realize that the world actually does go round. It can also be a help in realizing why an equatorial mount, which has an axis trained to point at the common centre of the star trails, will follow an individual star.

A simple camera on a firm mount can also be of some help in observing meteors when a shower is taking place. The camera will record both the trails of the meteors and the star trails. By recording times carefully, some accurate placing can be achieved.

With a fair-sized telescope, even an undriven one, one can also take good pictures of Venus, since this planet is so bright that an exposure of 1/25 second is enough with a moderate power eyepiece. All that is necessary is to focus the camera on infinity, and place it against the eyepiece, after the telescope has been focused by a person with normal eyesight.

With a driven telescope, an ordinary camera has some uses. It is easy enough to arrange a screw fastening on the end of the telescope, on to which the camera can be fixed. The telescope is then merely used as a drive for the camera. Very fair pictures of star fields can then be taken in this way. Unless, however, the drive is exceptionally accurate, one needs to watch some star in the field through a high power eyepiece. When, as it may do, it begins to drift, the slow motions are used to bring it back to the centre.

Still using an unadapted camera, some quite good work can be done by making a very simple bit of equipment. This consists of a piece of sheet metal bent at right angles. A hole is cut in one side, the same size as the outer thread of an eyepiece, which is then screwed into the draw-tube, holding the sheet metal. When so fixed, a camera can be placed on the tray, which is designed for size so that the centre of the eyepiece can rest against the centre of the camera lens. Once again, the telescope is focused accurately, and the camera set at infinity.

This is a somewhat elementary gadget; but it can supersede a purpose-built camera for special uses. For instance, few properly built cameras can be reduced to a sufficiently wide field to take in the whole of the moon. For photographing an eclipse, one needs to do this. I got a very good colour picture of an eclipse in this way, which would not have been possible with my more expensive and elaborate astronomical camera.

The next stage in elaboration is to make a camera specially for the telescope. One drawback of using an ordinary camera and an ordinary eyepiece is the number of lenses through which the light must pass, each sacrificing a little light, and each slightly degrading the final image.

There are various ingenious ways of adapting cameras for astronomical work; but this is not the place to discuss details. The two direct methods are to use either a plate camera with a ground-glass focusing screen, or else a reflex. Neither will make use of the camera lens, and, if a reflex is bought for adapting, it is enough to buy the body without the lens. In either case, the first thing is to fit an adaptor so that the camera body can be screwed on to the draw-tube in the same way as one screws in an eyepiece.

The simplest, and very cheap, way of securing a lens is to buy a single *concave* lens of such a size that it can be mounted in a tube which will slide inside the draw-tube. By sliding the tube to various distances from the primary image, but always on the far side of it, images of different sizes can be thrown back on to the plate or film and brought into sharp focus either with the ground-glass screen, if it is a plate camera, or in the viewer, if it is a reflex.

Quite good photographs can be taken by this simple method; but there is a severe limit to the size to which the image can be magnified. A fair example of this method is a picture of the moon showing half a diameter: that is to say, a quarter of the disc. It is not much use for planetary photography, since the developed size of the image will be very small indeed.

For closer pictures of the moon, or planetary photography, a *convex* lens is required. For those used to terrestial photography, a mental adjustment needs to be made. One must shed the idea that a large-diameter lens will give a better picture. A useful lens can be one from a small-sized cine-camera, which will only be a few millimetres in diameter. Since the image made by the telescope is so very small, a larger aperture lens is merely wasted.

A lens of this kind can be fitted into an inner tube, in the same way as suggested for a concave lens, bearing in mind two things. First, one is constructing an optical system similar to that of a magic lantern, and the lens must be reversed in direction to what it would be in the cine-camera. Secondly, the convex lens needs to be in front, that is on the camera side, of the primary image. The closer it is to the primary image, the greater will be the final magnification. By altering the position of the lens in the draw-tube one can increase or decrease the total magnification, and so also extend or diminish the size of the field which will be photographed. The cine lens I use myself on a telescope of 70″ focal length gives magnifications from about x30 to x80.

It is impossible to give any firm indication of the kinds of exposure in circumstances where there are so many variables. From Venus which, in proper circumstances, can be taken with an instantaneous exposure, the next step upward is the moon. This can also be photographed instantaneously with a 6″ reflector to a 12″, if the field is very wide – wide enough to take in the whole disc. For greater magnifications, depending upon the aperture of the telescope, the degree of magnification and the film speed, anything up to a second, or even over, may be needed. Jupiter, which is a difficult subject, but a rewarding one when successful, will need about four times the exposure of the moon in similar conditions. Saturn, which is a very difficult object indeed, will need at least twice the exposure of Jupiter. Mercury is almost impossible to photograph and the exposures needed for Mars vary greatly with how it is placed.

Perhaps it hardly needs stating that the astronomical photographer must either do all his own processing, or else use some friend who is willing to study the special technique. Very accurate timing of the developer is needed to get the best results. Moreover, the development process can be used to fiddle the exposure time. By cutting down the strength of the developer and extending the time of development, a slowish film may be used at the same rating as a nominally faster one. The great enemy of the astro-photographer is grain. Working from a tiny primary image, which for Venus, for instance, may be less than a tenth of an inch, one finishes with a photograph of five or six inches. Grain shows up badly on just the sort of subjects which are involved. Nearly all beginners start with great enthusiasm, and a film of very fast rating. The results soon procure a change of heart! The process of learning is the process of experimenting with all the variables; balancing length of exposure against the evils of grain; of small magnifications in the telescope against larger ones in the enlarging process. In astro-photography practice makes not perfection, but a reasonable satisfaction. The skill also calls for patience and enthusiasm. Apart from the moon, four or five good pictures represent a satisfactory year's work.

Another, and important, branch of astronomy – radio astronomy – involves instrumentation and techniques so wholly different from the rest of the science that anyone wishing to take the subject up really needs to make a quite separate study.

Radio astronomy, for all its difference, has become a tool of immense power and greater potential in cosmology, and has shed light on a whole range of astro-physical processes. As may so easily happen in any branch of science, this wide and fruitful skill owes its birth to an accident. Experimenting with radio transmissions, Karl Jansky, a communications engineer, discovered in the 1930s that certain background noise, which was a great nuisance, was not terrestial in its origin, but came from the Milky Way.

Looking back, there was no real ground for surprise. Radio

waves, after all, are just like any other electro-magnetic waves. They are emitted by matter in an excited state, and under the right conditions. Considering the vast range of physical conditions existing in stars, it is only to be expected that waves of all lengths should be produced, from the very short gamma rays, through visible light, and on into the radio wavelengths.

Once the search was started, discoveries came thick and fast and the whole gamut of the spectrum is now subject to observation by various methods. The first task was to design instruments capable of picking up faint signals in the radio wave band, and the huge dish telescope at Jodrell Bank, still the largest of its kind, was built through the force of character and vision of Sir Bernard Lovell.

Unfortunately, though the term is technically correct, the giant dish was called a telescope. The term is correct because, like an optical telescope, a large dish acts as does the mirror of a Newtonian, collecting electro-magnetic waves, and focusing them, but there the parallel stops. There was, a few years ago, an enterprising play on television, about a distinguished radio astronomer, who sat, night after night, looking through his radio telescope at sights no other living man had seen. Apart from the fact that even optical astronomers of high professional status seldom look through a telescope at all – they are mostly doing photography or spectroscopy – the radio telescope produces no visible image at all. The end product is a trace on a moving band of paper, which can subsequently be analysed in a computer.

Following the discovery that the sun emitted radio waves, a whole series of other celestial emitters were discovered, and the range of observations is still being constantly expanded. Among the earlier discoveries was a powerful source, which turned out to be coming from the Crab Nebula, the exploded super nova referred to in Chapter VII. Another, and somewhat unexpected source, was the planet Jupiter. Both the range of exploration and the distances of observation have been constantly expanding, until we can observe by the study of radio waves objects beyond the range of even the most powerful telescopes.

In order to resolve 'images', using the word in a wide sense, the receiver, be it a mirror or a radio receiver, size needs to be in proportion to the wavelength under observation. Since radio waves, though they themselves vary enormously in length, are nearly all vastly longer than visible light, the size of a radio telescope is required to be huge. Even the giant dish at Jodrell Bank proved to be much too small for some purposes, though its capacity to gather waves is quantitively very large.

Whereas Jodrell Bank is measured in feet, actually 250 feet, many radio telescopes are now measured in many hundreds of yards. They consist of lines of receivers, at angles to one another, which gives them an excellent resolution, though no very great 'light' grasp. Their disadvantage is that they are not, as the Jodrell Bank instrument is, steerable. Quite a number of radio telescopes are of the same nature as the few fixed optical telescopes, where one has to wait for the object which is to be observed to pass across the field of 'viewing'.

So far as the amateur is concerned, the bulk of observational techniques are far beyond his capacity. It is, however, possible for the amateur to do some useful work in this field. The cost of suitable equipment, oddly enough, lies far more in the recording devices than in the construction of the receiver, or telescope, itself. I once took the trouble to make a radio telescope for myself. Time, patience and application were called for in abundance; but, though the receiver was steerable, the actual cost was trifling. Honesty compels me to add that, though the instrument was excellent for picking up a television programme, and turned out to be quite a delicate recording voltmeter, as a radio telescope it was a complete failure.

The moral of this sad story is that the fault lay less in the design or construction than in my abysmal ignorance of electronics. Amateur radio astronomy is for the electronics expert who has astronomy as a side interest, rather than for the astronomer who wishes to extend into this new and rewarding field. Perhaps the best set-up is for a partnership between an electronics enthusiast and an astronomer. Even in professional astronomy, progress depends upon a series of partnerships. Observers, mathematicians, physicists and computer men must all work in closest harmony, together with a whole staff of technicians.

For the amateur who does wish to enter the field, there is plenty of information available, at various levels of expertise. *Practical Amateur Astronomy*, edited by Patrick Moore, has a chapter dealing with the matter; and there are a largish number of books on the subject for the amateur. It is only possible or relevant here to point out that, though it requires a wholly separate skill, radio astronomy is now, in some respects, almost as important as the other branches of the science.

Though it is hardly a branch of radio astronomy, there are other uses for radio waves. They form an essential part of the skill of astro-navigation, and a sort of radar is being used for survey work. Already the distance of both Venus and the moon have been determined with great accuracy by bouncing radio signals from them, and timing the period taken for the waves to go there and back. It is to this question of navigation to which we should now turn our attention.

CHAPTER XII

Navigation

Navigation, certainly as regards the older methods, is less a branch of astronomy than an application of it. From the earliest days of travel by sea, men found their way in a rough and crude way by observing the stars. The height of any star above the horizon at the time of local transit will, provided that one knows its declination, instantly allow the latitude to be determined. By a very simple extension of the observation, a position line can be found. Provided only that one knows the time accurately and is content to navigate by marking out a measurement on the surface of a globe, navigation is child's play. Columbus and other explorers of his era would have been delirious with joy if the method had been open to them.

We have noted previously that every star circles above the earth at a fixed declination. This is only another way of saying that it moves round at the zenith above a circle of latitude. Astronomical or nautical tables will tell one at a glance at which point on the latitude circle it will be at the zenith at any given moment.

If one is at that point at that time, the star will be at zenith. Now, if one is at some distance from that point, the star will be below the zenith. How far below will depend on how far one is away. Manifestly, for every degree, as measured from the centre of the earth, one is away from the zenith point, the star will be one degree below the zenith. Note too, that this is true irrespective of the direction of the zenith point from the point of observation.

Let us suppose that a certain star is, at a given time, observed to be ten degrees from the zenith, which is to say eighty

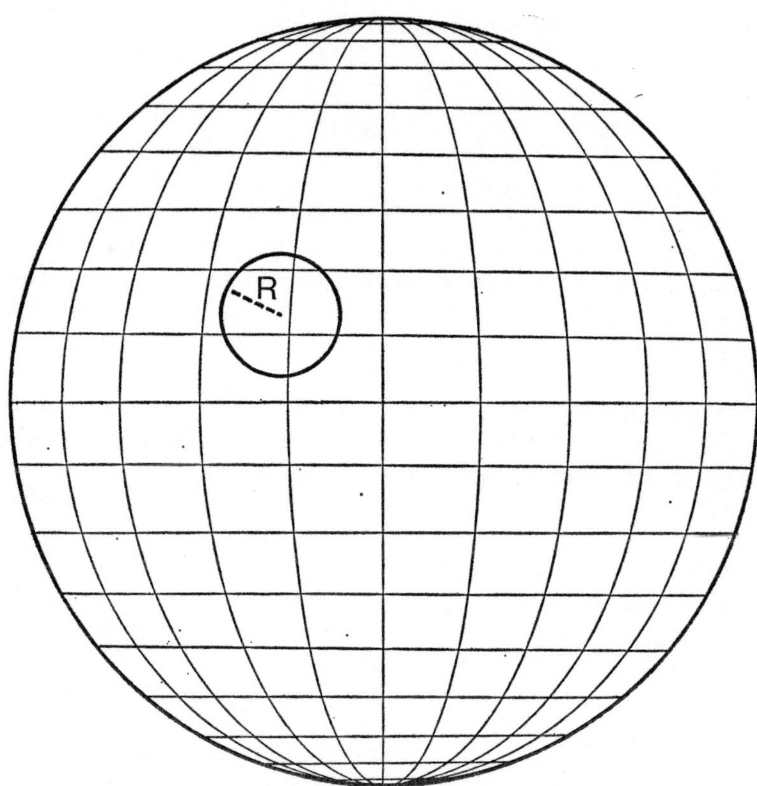

Fig. 47. The small circle is a circle of equal altitude. The radius
R is equivalent to the distance of the star observed from the zenith

degrees above the horizon as measured with a sextant. If one
looks up the zenith point for that star at that instant of time,
one may mark the position on the surface of a globe. If one now
takes a pair of compasses and opens them to a length equivalent
to a distance of ten degrees on the globe's surface, and draws
a circle on the globe, with the centre at the zenith point of
the star, it follows that the observer must be somewhere on
that circle. Measuring the ten degrees is most easily done by
putting one end of the compasses on the equator and stretching
the other to ten degrees of latitude. Be it noted that this length
represents ten degrees in any direction.

Having found the circle of equal altitude, as such a circle

is called, it is possible to proceed to measure the zenith distance of another star and repeat the process. Once again, a circle will be drawn representing all the possible positions of the observer. Naturally, the two circles will intersect. Where they cross one another is where the observer is. It is as simple as that.

Of course the circles will almost certainly intersect at two points, but dead reckoning will usually tell one at which of the two points one actually is. If one is so totally ignorant of one's position that this is not possible, one can measure the zenith position of a third star, when there will be only one point where the three circles intersect, giving one a uniquely determined position.

Navigation courses only frighten off the timid because navigation by drawing on the surface of a globe is not accurate enough for modern purposes. In order to transfer the circles which one could draw on the globe on to the flat surface of a chart, one requires to do a series of sums in spherical trigonometry. In any case, tables are now provided which obviate most of this tedious work, though navigators are, very properly, required to understand the fundamentals of navigation, as well as knowing where to look up what.

Naturally, and in practice, there are a certain number of extra factors to be taken into account, and limitations to the observations which can be made. For instance, the altitude of the star, as observed from the deck of a ship, will have to be corrected for the height of the eye above the horizon – though this can be avoided by using a bubble sextant, as is done in a plane. Then, and especially if the star is anywhere near the horizon, there must be an allowance for refraction: the bending of the light rays through the atmosphere, which will make a low star appear higher than it actually is.

A more severe limitation is imposed by the need to be able to see both the star and the horizon at the same time. This generally limits a plot to just before dawn or just after sunset, and also necessitates using a star bright enough to be observed in the twilight. It will not often be possible to take plots from two stars in these conditions. Sophisticated equipment will overcome this limitation. Electronics are so advanced that it is possible to have a telescope which can be locked on to a fourth magnitude star in broad daylight!

In general, a good deal of astro-navigation is done by using the sun rather than a star. The zenith position of the sun, of course, varies in declination throughout the year; but its position at any given moment can be abstracted from the tables. Taken at local noon – the familiar noon sight – gives an instant determination of latitude, from which dead reckoning can be corrected in one direction. At other times, the sun provides, as does a star, a circle of equal altitude, on which the observer must lie. For practical purposes, this is good enough by itself as a check on dead-reckoning. If necessary, two solar circles can be found at a fair interval, and the only allowance for an absolute fix is the distance and direction of the run between the two observations.

With a good many differences, the measurement of stellar angles will be employed in the most modern need for navigational information – space travel. Another modern navigational method is already used extensively in the guidance of space probes; this is called inertial guidance, which is also used, for instance, in submarines which are submerged for long periods.

We spoke earlier of inertia as a natural laziness of matter, and the principle is used, in a crude form, for an everyday purpose; the measuring of the efficiency of the brakes of a car. If one has a weight which is allowed to move freely on rollers against the pressure of a spring, one has an accelerometer. If placed on the floor of the car, the weight will try to go on forwards when the car is braked. The more severe the braking, the harder the weight will press against the spring, and the further it will move a needle against a scale, so measuring the deceleration due to braking. If the gadget is placed the other way round, the weight will hang back as the car accelerates. Once again, the rate of acceleration can be measured.

If one integrates acceleration with time, the result is velocity. If that too is integrated with time, the result is distance travelled. The double integration opens the door to severe and cumulative error, and only a fantastically accurate instrument will be at all serviceable. In a space probe there is further complication. Movements can take place in three different directions, along the axes one may roughly call length, breadth and height. All three measurements need to be made and recorded. Fortunately modern technology is equal to the challenge.

In place of a weight, a modern inertial guidance system uses a series of gyroscopes. If one spins a gyroscope, and then accelerates it, the result will be that the gyroscope will begin to precess. A feed-back control is fitted to prevent the precession, and the force needed to do this is the measure of the acceleration. Since a gyroscope has the property of resisting a change of direction, the system also keeps the guidance instrument stable as the probe, plane or submarine alters direction.

Making gyroscopes of the needed accuracy is no mean feat. The parts are machined under a microscope, in a room rigidly controlled for temperature, and in which the least speck of dust has been filtered out. Even the machine operators have to wear masks.

If one might, perhaps, say that inertial guidance is the navigation of the future and astro-navigation the system of the past, the present basis of navigation is largely electronic. There are so many radio navigational systems that, in a short summary of this sort, is hard to know where to begin – or where to end! Radar is the simplest of the methods used. In this, as mentioned, we only have to send out a radar pulse, and wait to get the echo. The principle may be used either, as in ships and planes, to observe and display obstacles in the path of the observer or, by timing the interval before the echo arrives, to calculate the distance. A variation of this method is coming into use, whereby a beam from the receiving end is employed to trigger a transponder on the plane. This causes it to send out a message, already coded and waiting transmission. Thus, in approach to an airport, where radio channels are already heavily employed, an incoming plane may record automatically the details of its course and position, which the ground controller can obtain by merely sending a signal and getting an instant response.

An even simpler and earlier system consists of a mere set of radio beacons. Their respective directions may be found by using a rotating directional aerial and a plot made on a chart. A variation of this is a system in which each beacon transmits different signals in different fixed directions. By listening to a signal from such a beacon, with an ordinary radio set, the navigator can tell at once, by the coded signal he receives, in what direction the beacon lies. Three such beacons will give a fix, though not a very precise one.

More complex radio systems rely upon the process known as phase comparison, the details of which are beyond our scope here. If one thinks of a radio wave as much like a wave in the sea, it will consist of a series of crests and troughs, though with radio waves it is a matter of ebb and flow of potential; but the simile is adequate for our purposes.

Let us suppose that we have two transmitters, some distance apart and so linked that they emit crests and troughs in step with one another. If the wavelength is long, say just over a kilometre, anyone picking up their signals will receive one crest from each station about every three microseconds. If crests from each of the two are received simultaneously, then the receiver must be the same distance from each station, or else a distance different by whole multiples of the wavelength. If a crest from one appears at the same time as a trough from the other, the difference in distance of the observer must be half a wavelength.

Now lines of equal difference of distance take the form of hyperbolae, as shown in Fig. 48. Let us suppose that the lines each represent a difference of three microseconds, or one wavelength. If the receiver is getting crests and troughs simultaneously, he will know that he is somewhere on one of these lines, though he will not know how far along, nor on which line. Suppose again, however, that there is another pair of sending stations, in line at approximately right-angles to the first pair, from which the signals are being received, the hyperbole lines will cross, forming a lattice work of lines.

At the start of a journey, the navigator will know his position exactly and can mark it on a chart which shows the hyperbole lattice. An instrument will show him every time he crosses one of the lines coming from either pair of transmitters. Since he knew his original position, he will be able to mark a fresh position every time a line is crossed. In practice, the instrument will keep track not only of whole multiples of a wavelength, but continuously of fractions. By referring to dials, the navigator will have an exact fix available at any moment.

Developments of this system have gone so far that a child could now keep track of a fast-moving vehicle. The Decca system is based on these principles, and has become delightfully simple, accurate and easy. To begin with, the system, in addition to the dials, offered an alternative of a pen recorder which kept a con-

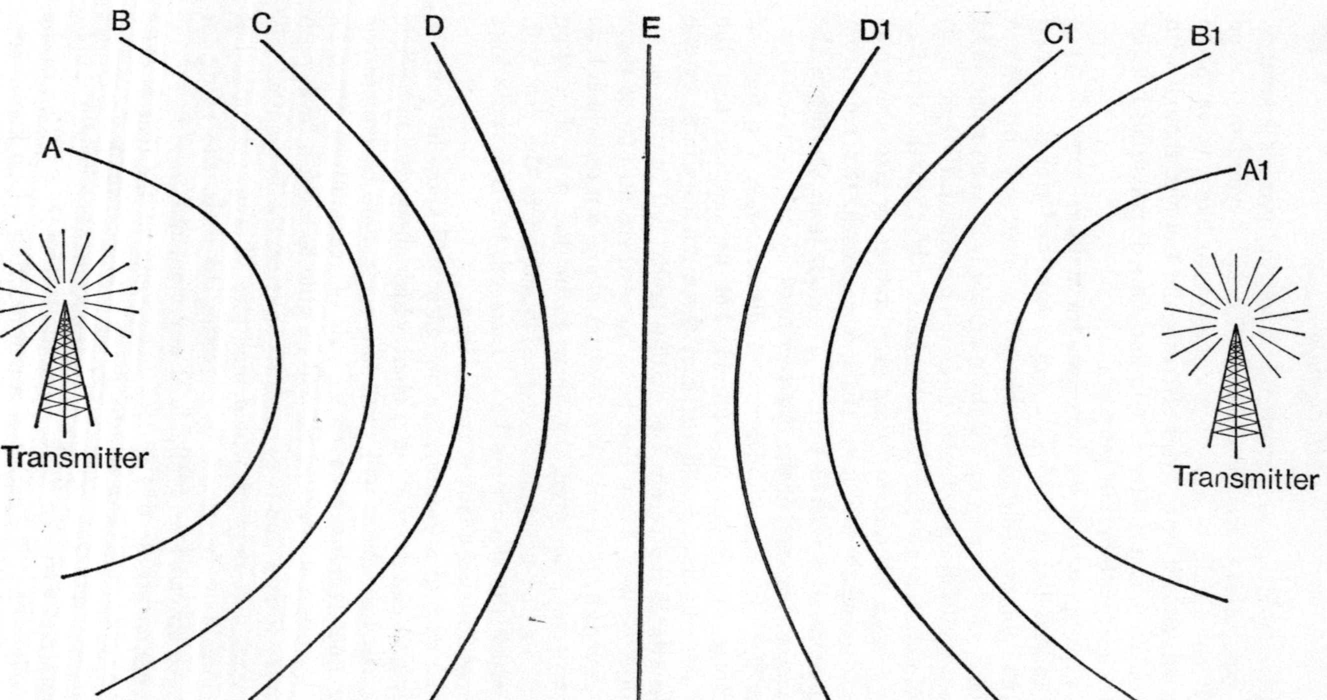

Fig. 48.   If the signals from the two beacons are received simultaneously, the observer is on the line E. If he is a given and fixed difference in distance from the two transmitters, he will be on D or D1. If the difference is slightly greater, he will be on C or C1, and so on

tinuous trace on a chart. The chart had to be deformed to represent the curved hyperhole lines as straight. The result was a very deformed chart indeed. A further development has overcome this handicap, and the pen now traces out the movement of the ship or plane on an ordinary chart. A single glance will show the navigator his exact position. In a plane coming into land even the time taken to read dials, and then plot a position on a chart, might be too much. To have one's position constantly recorded on a standard chart or map represents a great triumph, and a great boon.

To astronomers, these interesting developments are of some indirect concern, since probes moving far out in space will need all the navigational aids which the ingenuity of man can devise. Separate transmitting stations on the moon may be an early requirement; and it does not need a far stretch of imagination to suppose that, before so very long, there may be others on Mars. All the main methods we have referred to may have their place, in adapted forms, in the navigation of the future, together with others which would have no function on our own planet. It is not too much to suppose, for instance, that when probes are accelerated to very great speed, the Doppler shift of radio signals could be used to measure velocity.

Just as the Decca pen recording system would, only a few decades ago, have seemed a crazy dream, we may see other developments of which we have as yet no conception. The man of the space age must learn that the impossible of yesterday is the project of today and the achievement of tomorrow.

CHAPTER XIII

Journey into Space

Though, perhaps, a hundred years ago space travel would have seemed – did seem – the wildest flight of fancy, the strange thing is that unlike, for instance, radio communications, the bar to its achievement was a matter not of lack of scientific knowledge but of technology. The principle of the rocket was known a very long time ago. During the Napoleonic wars. Congreve's rockets had a spectacular, if erratic, career. Unlike other means of propulsion, the rocket needs no air to maintain its flight. It can work in a vacuum rather better than in the atmosphere, since there is no air resistance. To travel away from our own planet by rocket propulsion presents no difficulty in principle. The requirement is simply to construct a rocket with enough fuel to provide it with the required velocity, and yet not be so heavy that the take-off weight exceeds the thrust.

The principle of the rocket is easily demonstrated. If one sits on a swing and throws a heavy weight forward, and away from one, the result will be that the swing will move backwards. The principle is easily stated as: action and reaction are equal and opposite. If one multiplies the mass of the object thrown by its velocity, it will be equal to one's own weight, plus that of the swing, multiplied by one's velocity backwards. The opposite effect can be achieved by throwing another similar mass forward as one arrives forwards at the bottom of the swing. The impetus imparted is now backwards, and will stop the swing from moving. The first process is equivalent to firing a rocket at rest to impart motion, the latter to firing a retro-rocket to stop motion. A rocket does not throw weights about; it fires hot gases. But a gas, like any other form of matter, has mass and the effect is the same. By firing the gas backwards, one propels the rocket forwards. A rocket, unlike a jet engine, which relies upon air in which to burn its fuel, carries its own supply of both fuel and oxidant, and so is independent of the air. The firework rocket burns gunpowder; but space rockets use all kinds of fuel and oxidant, kept separate until the moment of combustion.

As we have noted, the first consideration is that the thrust at the moment of firing must be more than the weight of the loaded rocket; but it need be only a little more, since, as the fuel burns away, the load is decreased and the acceleration imparted consequently increases. This is why a rocket, which will attain fantastic speeds, starts off slowly.

These basic facts were known to people in the past, who dreamed of space travel and were written off as cranks and madmen for their vision. One such 'dreamer' was Konstantin Tsiolkovsky, a Russian, born in 1857, who wrote on the subject, and who even designed a rocket which bore no small resemblance to the early satellites and probes. Less scientifically, but more vividly, Jules Verne and, later, H. G. Wells, amongst others, wrote stories of space travel, though the former made the mistake of having his vehicle fired from a gun. The Second World War gave the science a great impetus, with the construction of the V2 weapons; enormous rockets containing explosive heads. Many of the German scientists who worked on this project emigrated after the war to Russia or America, and carried on from destruction to scientific exploration. Notable amongst these was Wernher von Braun, who has been the main architect of the highly successful U.S. space travel programme.

Here we must concern ourselves less with the history of astronautics than with the basic principles involved, though one must pause to remember the two great days in the history of man. The first in 1957 when the Russians sent up the first artificial satellite, and the second in 1969, when two Americans landed on the moon and were later returned safely back to earth. In the whole of human history there have been no achievements to compare with these landmarks. The somewhat overrated voyage of Columbus, when he first discovered the West Indies, is often cited as a parallel; the comparison, however, is but a very feeble one.

Naturally, the distance which one wishes to travel in space dictates the velocity at which the rocket will be required to travel. It is only necessary to think about throwing stones straight up. The harder one throws them the higher they will go. If one threw hard enough, or was able to throw hard enough, the stone might hit the moon – in theory, at any rate, In practice, it would be moving so fast that it would be burned up in the earth's dense lower atmosphere soon after being thrown.

The limiting necessary velocity is the velocity at which the object sent up would never return at all. The concept is a little elusive. No matter how far one travels away from the earth, there is always theoretically a residual gravitational attraction from the earth. The attraction gets feebler and feebler as one moves away, but it is never zero. On the other hand, the less the gravitational attraction the less the velocity required. The two quantities diminish simultaneously. Escape velocity, as it is called, is the initial velocity needed, so that the residual required velocity is always slightly less than that needed to overcome the reduced gravitational attraction. They will reach zero together at infinity. This, of course, ignoring the gravitational fields of other bodies.

To calculate this escape velocity for any planet, or other starting point, one can reverse the problem, and enquire what would be the velocity of an object, coming from infinity, when it reached the surface of the planet. The acceleration, due to the gravitational field, would start at zero and finish at the value obtaining on the surface of the planet. In the case of the earth, the terminal acceleration would be 981 cm. per sec. per sec. The total velocity, in these circumstance, can be calculated by using a very elementary form of integral calculus, as is shown in the Appendix.

The escape velocity for any body is the square root of twice the surface gravity multiplied by the radius of the body. In the case of the earth this works out at about eleven and a quarter kilometres a second. Any probe leaving the earth will need something nearing this initial velocity. If we were to be dealing with a stationary earth and moon, with no other gravitational bodies to confuse the issue, we should only need to carry a moon probe to the point where the moon's gravitational attraction became stronger than that of the earth. The initial velocity, however, would still be nearing the figure for escape velocity. In practice, we have to consider a whole series of factors, including the movement of the earth around the sun, the gravitational attractions of the sun, as well as that of earth and moon, and the relative motions of the earth and moon. Probes move in tracks very far removed from straight lines.

With infinite amounts of power available, it would be theoretically possible for a probe to set off in a straight line, and continue in one, from the point of departure to the point which would be reached at the moon, or other target. Such a procedure, however, would entail continuous changes of course, and continuous expenditure of power. This would be needed not only because the target itself is moving, a difficulty which would be overcome by good aiming, and not only because the constantly changing gravitational fields would cause readjustments, but also because of the differing orbits of the earth and the target.

The point can be illustrated by what is called the Coriolis effect, which causes winds round depressions in the northern hemisphere to move anti-clockwise, and in the southern hemisphere clockwise. A volume of air at the equator, which is 'at rest', is, in fact, moving to the east with a speed of around a thousand miles an hour; that is to say with the speed there of the earth's rotation. At the pole, the air will be more truly stationary. When a depression north of the equator sucks air up from the equator, the air will naturally have a slant to the east. Air coming from the pole, on the other hand, will tend to be 'left behind' by the land over which it flows, and will have a westward slant.

The same effect can be used to persuade oneself that the earth is in fact rotating, or to pose a nasty question to a disciple of the flat-earth movement. Suppose one built a high tower on the equator – though it would have to be a very high tower indeed! – and dropped a stone. What would happen to it? If the tower were high enough, and narrow enough, the stone, instead of landing on the floor, would hit the wall on the eastward side.

The reason is easy enough to appreciate. The top of the tower would be moving faster than the bottom, because the rotation time would clearly be the same as the bottom, but the circumference of the circle through which it moves would be greater. It would have further to travel in the same time. In fact, with delicate equipment and careful measurement, the experiment can be, and has been, performed.

The same sort of effect is of importance in space navigation. The planets are moving in different orbits round the sun, and at different speeds. A probe moving from one orbit to another will have to take into account factors such as this in selecting the most economic path. It is fairly easy to see the principle at work if one takes into consideration the behaviour of a weight swung round on the end of a piece of elastic.

A planet, as we saw earlier, is in such an orbit that the centrifugal 'force' is precisely balanced by the gravitational attraction of the sun. If it could be accelerated, it would go into a wider orbit. If it were slowed down by retro-rockets, it would fall into a smaller orbit. There are differences from a weight on the end of the elastic; but the results are the same. Suppose then, that one is swinging such a weight round, and one wishes to move the weight out to a larger orbit, one merely increases the speed of rotation, and out goes the weight.

If one took a series of rapid exposures of the weight altering orbit, it would be seen that the transfer from orbit to orbit was not made by a simple jump from one to the other. The weight would move in a curve, as shown in Fig. 49. This, in practice, is how a probe moves from the the earth to some other planet. Remembering that both the earth and the other planets, not to mention the moon, are all orbiting round the sun, one can reduce the transfer of orbits to a process similar to that of shifting the orbit of the weight on the end of the elastic. If one wishes to go to a planet in a wider orbit than our own, it can be done by a simple acceleration. If to a lower orbit, then by retardation. Naturally, it is not as simple as all that; but the principle is the same. Transfer orbits, as they are called, are basically achieved by speeding up or slowing down the motion of the probe in relation to the solar orbit.

Some of the complications become obvious if we think about the matter for a while. Let us suppose, for instance, that we have a circular racetrack for cars, and that there are two cars going round the circuit at the same speed, but at opposite sides. If the driver of one car wishes to catch the other, all he has to

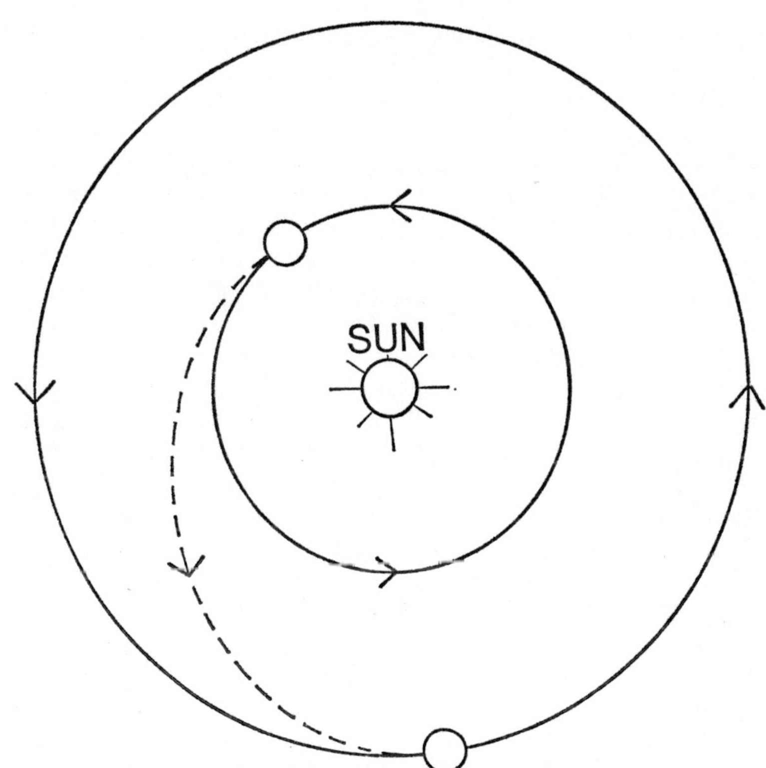

Fig. 49. The dotted line shows a transfer orbit from a nearer planet
to a further one

do, if it possesses the necessary acceleration, is to speed up his
car. In due course, and if that is what he wishes, he can run
into the back of the car he is trying to overtake.

One can imagine exactly the same situation with two satellites
in circular orbits round the earth. But, though the laws of
motion are held to be exactly the same, the procedures would
have to be quite different. If the overtaking satellite were merely
to accelerate, far from catching up the target satellite, it would
be thrown into quite a different orbit. Depending upon the
degree of acceleration, and the period over which the rockets
were fired, the overtaking satellite would be put into an elliptical orbit, of a shape determined by those factors.

It is hard for mortals, reared in a particular gravitational field, and limited by air resistance and friction, to realize how much the same laws of physics lead to totally different results in different conditions. Let us take the motor racetrack we were talking about, and transfer it to the moon: the results of endeavouring to run a race would be entertaining. The mass of the cars, and therefore, the inertial resistance to change of motion would be the same, but the gravitational force only about one sixth. Even supposing that the tyres could be made to grip, any effort to pursue a circular course would result in vicious skidding, because the centrifugal force would be the same on earth, but the frictional resistance far less. Adding weight would be useless; acceleration would be sacrificed and centrifugal force increased in proportion to the frictional resistance. On a giant planet, the position would be reversed. Leaving out of account what would happen to the driver, the car would collapse under its own weight. The fact is that this earth on which we live is a peculiar place, in that the conditions in which life, as we know it, can flourish, are limited to a very narrow range, which just happen to obtain here. True, we could live without motor cars – we might, indeed, be a lot happier and healthier – but we can only flourish in a narrow range of temperatures out of the vast range to be found in nature; we can only breathe in an atmosphere with enough oxygen and in the absence of a whole range of other gases; we cannot stand up to radiations from the sun, from which our atmosphere protects us ... The limitations are almost endless, and yet our earth contrives to meet the exacting conditions required. When we take to space, we also enter upon hazards, not all of which are even known, and some may be unguessed.

One of the rather unexpected hazards, thinking about the matter *a priori,* is the effect of prolonged 'weightlessness'. Concerning this state, there is a very persistent misunderstanding. It is widely believed among laymen that the state is the consequence of passing out of the earth's gravitational field. But, as we have seen, we can never do this, let alone shaking off the gravitational fields of other bodies. 'Weightlessness' is a condition which we can experience on earth, as easily as in space, and is achieved in practice, as part of an astronaut's training. We are not weightless at all, but in a state of free-fall.

If one were to evacuate the air from a mine-shaft,* and then jump down it, we should be in a state of free-fall or 'weightlessness', until we hit the bottom of the shaft – when the experiment would terminate abruptly. Less drastically, the same experience can be achieved by putting a plane into a course which follows that which would be taken by a body inside it in the absence of air and given its commencing velocity. This is precisely what is happening to a probe or satellite, be it natural or artificial, when it is not being artificially accelerated. It, and anyone on or inside it, is yielding itself without resistance to whatever gravitational forces are acting upon it. The feeling of weight, on the surface of the earth, is due to the fact that we are resisting the pull of gravity; as we are caused to do by our contact with the surface upon which we stand.

The future of space travel is a subject upon which it is difficult to write, because so many prophecies have been falsified by events. The most fruitful subject for popular imagining is the possibility of getting altogether away from our immediate environment, out into the depths of space, or of others from afar coming to visit us.

The limitations upon really distant travel are many and complex. The first is the need to have food, drink and air; but this is not so severe a limitation as it appears. Given enough power, it could be possible to set up a cycle in which all the wastes of the human body could be collected and reprocessed, and used over and over again. It is the power which is the first real limitation.

No matter how technology may be advanced, there is a theoretical limit to the power which can be got from chemical combustion. Real long-distance travel will almost certainly have to wait until more basic processes can be harnessed, and harnessed without equipment of great bulk and weight. Clearly, the ultimate will be when we can destroy matter altogether, or rather transmute it into energy. In the less-remote future, there are some avenues of possible progress. One, much discussed, is the idea of firing electrons backwards. The amount of force derived would be trifling; but it could be maintained for very long periods. Since the acceleration is a function of both time and force, and efficiency dependent upon the velocity

---

* Merely to obviate the minor factor of air friction.

of ejection, such a system could be very effective. It would still, however, be trifling in comparison with a whole matter transmutation.

The most serious limit to travel is the velocity of light. It is basic to the theory of relativity that nothing can travel faster than light. The more rapidly a body moves, the greater becomes its mass. At the sort of velocities with which we deal, this can be quite ignored; but, as velocity increases, it becomes important and then dominant. At the velocity of light itself, the mass would become infinite. At the moment, and for inter-planetary travel, this factor is still negligible; but when we start to think about moving even about our local galaxy, the matter is very different. Even at the velocity of light, it would take a hundred thousand years to cross it from end to end. To move outside we start to deal in millions of years, and then in hundreds and finally thousands of millions. Let us too remember that a probe starts off at about eleven kilometres a second while the velocity of light is three hundred thousand!

One may set the limit where one likes, according to the degree of vividness of one's imagination; but a limit there must be some-where. Human ambition may be boundless, and imagination free to soar where it likes. The hard fact is that the immensity of our cosmos is such that though our achievements may be wonderful, our horizons still ever expanding, man is nothing but the pygmy he always seemed in the vastness of creation.

Astronomy can be a study of both wonder and satisfaction; but it is a discipline which teaches us at one and the same time how magnificent are our powers of thought and speculation; but, in the final essence, how insignificant we are alike in time and space.

Appendix

I have made use in this book of the concept of centrifugal force, in spite of the fact that it is only a useful fiction, much disliked by scientists. It is, however, regularly used by engineers, is simple to understand, and gives the right answers. I have used the concept, not only because it is simple, but also because it is familiar to everybody. If we swing a weight round on the end of a string, we can feel the tug, tending to pull our hands outward. A spring scale jointed into the string will register the pull, which increases with the speed of rotation.

Centrifugal force is a fiction, as we can demonstrate for ourselves. If we swing the weight on a piece of cotton so hard that the cotton breaks the weight will not fly outward along what was the line of the cotton, as we might be tempted to expect it to do it there were a real force in that direction. It would not even fly outward and forward, as it would really do if there were such a force. It would fly off at a tangent, as in one of the arrows in Fig. 7. (page 23)

If the effect had been called a centripetal force, that is a force acting inward, there would be less objection. The tug we feel is really an *inward* tug, supplied by the hand, and which is causing the weight to deviate from the straight line along a tangent, which, but for the tug, it would take. It is, therefore, really the name only which is objectionable; but it would only add to the confusion if I were to refer to centripetal force where references in other books were to centrifugal force.

It is the most basic of the laws of motion that any object will move in a straight line unless and until some force is applied to it which will cause it to alter either its direction or its rate of motion. A circle can be thought of as an infinite series of infinitely small tangents – a regular figure with an infinite number of sides. When our weight is swinging round on the string, it is, for an infinitely short time, travelling in a straight line. The tug of our hand inward causes it to alter direction continuously. The tug, in fact, is an inward force applied by our hand to make the weight pursue a circular course when, otherwise, it would shoot off in a straight line.

When we come to Newton's calculation showing that the moon obeyed the same laws of motion as applied on earth, we can explain it in either of two ways. We can say that the centrifugal force of the moon's orbit is exactly balanced by the gravitational attraction of the earth. Most laymen and beginners find this a simple and easy proposition to follow. The alternative explanation is that the moon is in free fall, and that it is in a state of constant acceleration towards the earth, to which it never gets any closer. Most beginners find this statement incredible and self contradictory; but, in truth, it is the more scientific statement.

For these purposes, we are neglecting the small amount of the earth's movement round the barycentre, and also ignoring the slight eccentricity of the moon's orbit, taking the radius of rotation as the moon's average distance from the earth.

If we take a very short piece of the moon's orbit and treat it as consisting of two separate motions, one in a straight line along the tangent, and the other its acceleration under the force of gravitation towards the earth's centre, we can easily work out where it will be after say ten seconds. It will have gone so far along the tangent in a straight line, and also so far towards the earth's centre. If we plot this position, we shall find that it is very near to being on the circular orbit. If we repeat the process from this new point, working out the distance along the new tangent, and towards the earth's centre, we shall again be near to the circular orbit. The smaller the distance we take, the closer to the true orbit it will be. If we make the distances infinitely short, or rather if we could do so, the successive points would lie exactly on the circular orbit. If this sounds like Alice and the red Queen, who had to run as hard as they could to stay in the same place, one needs to remember that the accelerations change direction constantly. The acceleration towards the earth at any given moment is in exactly the opposite direction in space 180° later, though still towards the centre of the earth.

Using the centrifugal force principle it is easy to see that the attraction of the earth on the moon will exactly balance the centrifugal force, which is given by the equation: $c.f. = mv^2/r$, where m is the mass v the velocity and r the radius of rotation. If we consider a unit mass at the centre of the moon, the m disappears, being unity, and we have $cf = v^2/r$. The velocity is merely the distance covered in a single rotation in the time taken to do it. That is $v = 2\pi r/t$, where t is the time taken. Substituting, we get $c.f. = 4\pi^2 r/t^2$. For rough calculation we can take π^2 as 10, and say $cf = 40r/t^2$. r will be the distance in centimetres from the centre of the earth to the centre of the moon: 3.844×10^{12}. t, the time of one lunar revolution, is 2.360×10^6 seconds. The force therefore amounts to .28. To calculate the gravitational pull on a unit mass at the centre of the moon very approximately we can take the moon's distance as 60 times the radius of the earth. As gravitational attraction falls off as the square of the distance, the attraction at the centre of the moon would be $1/60^2$ of the attraction at the earth's surface (i.e. 980 cm. sec.^{2}), or .27. Considering the huge numbers we are using, and the rough approximations we have used, that is very fair agreement. Anyone who wishes to do the calculation more accurately will only need to repeat the sum using exact distances.

Escape velocity

In coming from infinity to the surface of the earth or other body, any mass would lose potential energy and gain kinetic energy. The amounts of the two would be equal. The potential energy lost is the mass of the object multiplied by the distance travelled and the gravitational attraction. If x is the distance, at any point in the journey to the surface of the earth or other planet, and the gravitational attraction on a unit mass at that

point will be $\dfrac{MG}{r^2}$ where M is the mass of the earth or other

body, and G the constant of gravitation. The potential energy lost, therefore, will be the sum of all the values of x from infinity to the surface of the body, which will be distant by a radius r from its centre. We get:

$$\int_{r}^{\infty} \frac{MG}{x^2}\, dx = \frac{MG}{r}$$

This must be equal to the kinetic energy of the falling mass, which is given by $\frac{1}{2}mv^2$ with m as unity. So:

$$\frac{MG}{r} = \frac{1}{2}v^2$$

But $\frac{MG}{x^2}$ is the attraction on a unit mass at the surface of earth or other planet and can be called g. Therefore

$v^2 = 2gr$ and $v = \sqrt{2gr}$

Escape velocity for any body is therefore the square root of twice the surface gravitation multiplied by the radius of the body.

To use this equation, one needs to know the surface gravity of the body in question as well as the radius. If the mass is known, one can substitute $\frac{MG}{r^2}$ for g, thus getting:

$$\text{ESCAPE VELOCITY} = \sqrt{2\frac{MG}{r^2} \times r} = \sqrt{\frac{2MG}{r}}$$

If the specific gravity of the body is known, an easy calculation can be obtained from the equation:

$$\text{ESCAPE VELOCITY} = r \times \sqrt{s} \times 7{\cdot}5 \times 10^{-4}$$

where s = the specific gravity, and r the radius in centimetres. The answer will be in centimetres per second.

INDEX